The Airliner Cabin Environment

Air Quality and Safety

Committee on Airliner Cabin Air Quality
Board on Environmental Studies and Toxicology
Commission on Life Sciences
National Research Council

NATIONAL ACADEMY PRESS
Washington, D.C. 1986

NOTICE: The project that is the subject of this report was approved by the Governing Board of the National Research Council, whose members are drawn from the councils of the National Academy of Sciences, the National Academy of Engineering, and the Institute of Medicine. The members of the committee responsible for the report were chosen for their special competences and with regard for appropriate balance.

This report has been reviewed by a group other than the authors according to procedures approved by a Report Review Committee consisting of members of the National Academy of Sciences, the National Academy of Engineering, and the Institute of Medicine.

The National Research Council was established by the National Academy of Sciences in 1916 to associate the broad community of science and technology with the Academy's purposes of furthering knowledge and of advising the federal government. The Council operates in accordance with general policies determined by the Academy under the authority of its congressional charter of 1863, which establishes the Academy as a private, nonprofit, self-governing membership corporation. The Council has become the principal operating agency of both the National Academy of Sciences and the National Academy of Engineering in the conduct of their services to the government, the public, and the scientific and engineering communities. It is administered jointly by both Academies and the Institute of Medicine. The National Academy of Engineering and the Institute of Medicine were established in 1964 and 1970, respectively, under the charter of the National Academy of Sciences.

This project was prepared under Contract No. DTFA01-85-C-00013 between the National Academy of Sciences and the U.S. Department of Transportation.

Available from: National Academy Press, 2101 Constitution Avenue, NW, Washington, D.C. 20418
International Standard Book Number 0-309-03690-9

Printed in the United States of America
First Printing, August 1986
Second Printing, November 1986
Third Printing, March 1987
Fourth Printing, March 1988

COMMITTEE ON AIRLINER CABIN AIR QUALITY

Thomas C. Chalmers (Chairman), Mount Sinai Medical Center, New York, New York
Paul F. Halfpenny (Vice-Chairman), Van Nuys, California
John D. Spengler (Vice-Chairman), Harvard School of Public Health, Boston, Massachusetts
Harriet A. Burge, University of Michigan, Ann Arbor, Michigan
Edward J. Calabrese, University of Massachusetts, Amherst, Massachusetts
Stephen M. Horvath, University of California, Santa Barbara, California
Nan M. Laird, Harvard School of Public Health, Boston, Massachusetts
* Genevieve Matanoski, Johns Hopkins University, Baltimore, Maryland
James M. Melius, National Institute for Occupational Safety and Health, Cincinnati, Ohio
Matthew I. Radnofsky, Seabrook, Texas
Donald H. Stedman, University of Denver, Denver, Colorado

National Research Council Staff

James A. Frazier, Project Director
Devra Lee Davis, Acting Director, BEST
Andrew M. Pope, Staff Officer
Rob Coppock, Staff Officer
Alison Kamat, Information Specialist
Judy Tiger, Administrative Secretary
† Jean Perrin, Administrative Secretary
Norman Grossblatt, Editor

* Until February 1986.

† Until December 1985.

BOARD ON ENVIRONMENTAL STUDIES AND TOXICOLOGY

Ex Officio

Gary P. Carlson, Purdue University, Lafayette, Indiana
Thomas Chalmers, Mount Sinai Medical Center, New York, New York
Arthur B. DuBois, John B. Pierce Foundation Laboratory, New Haven, Connecticut
Alan M. Goldberg, Johns Hopkins University, Baltimore, Maryland
Bernard D. Goldstein, Robert Wood Johnson Medical School, Piscataway, New Jersey
David Jollow, Medical University of South Carolina Charleston, South Carolina
Roger O. McClellan, Lovelace Inhalation Toxicology Research Institute, Albuquerque, New Mexico
Norton Nelson, New York University Medical Center, New York, New York
Duncan T. Patten, Arizona State University, Tempe, Arizona

National Research Council Staff

Devra Lee Davis, Acting Director, BEST
Jacqueline Prince, Staff Associate

Preface

This study came about because a series of Congressional hearings in 1983 and 1984 revealed that the available data on airliner cabin air quality were contradictory. Concern was expressed about the absence of standards for many aspects of cabin air quality that annoyed passengers and crew. The regulatory community and the airline industry asserted that present standards and practices were adequate and that the aircraft environment endangered the health and safety of neither passengers nor crew.

As a result of the hearings, Congress, in Public Law 98-466, mandated that the National Academy of Sciences conduct a study to determine whether air quality and standards aboard commercial aircraft are adequate for the health and safety of all who fly. The Academy was asked to determine whether such aspects of cabin air as the quantity of outside air, the quality of onboard air, the extent of pressurization, the characteristics of humidification, the presence of cosmic radiation, contaminants (such as bacteria, fungi, and other microorganisms), and pollutants (such as environmental tobacco smoke, carbon monoxide, carbon dioxide, and ozone) could be responsible for health problems in the long or short run; to recommend remedies for problems discovered; and to outline the safety precautions necessary to protect passengers in event of in-flight fires, which produce smoke and fumes. Accordingly, the Committee on Airliner Cabin Air Quality was established in the National Research Council's Commission on Life Sciences.

Issues the Committee addressed included the following: Are there problems with the air quality in commercial airliners? If so, what is the potential public health significance for those exposed over the short or long term? Are the problems solely those of brief discomfort, or are the health and safety of crew and passengers threatened? How well established is the threat? What can be done to alleviate it?

The Committee has reviewed the available pertinent information to reach an independent scientific consensus on these issues. Unfortunately, evidence on these questions is sparse, especially on health effects. Carefully designed epidemiologic studies of health effects associated with air travel are virtually nonexistent, and most of the relevant published reports deal only with specific incidents. Hence, it is difficult to evaluate the risk to the exposed population. Indeed, the dearth of pertinent data limits conclusions about the potential for adverse health effects to no more than estimates. Much more research must be conducted before risks can be accurately assessed.

The words "health" and "safety" are emphasized throughout. The Committee found it difficult to pigeonhole problems neatly as related to health, safety, comfort, or combinations of these. For example, the time required to evacuate a plane if fire occurs is certainly a safety issue, but it is also a health matter, in that evacuees will be subject to toxic fumes for a longer or shorter time. Cigarette-smoking might be primarily a comfort issue for both nonsmokers exposed to smoke and smokers deprived of their stimulant; it might also be a health issue for nonsmokers, as well as smokers; it is certainly a safety issue if cigarettes are improperly disposed.

The importance of these distinctions is that the Federal Aviation Administration (FAA), for which this study was prepared, might not have the statutory authority to deal with some issues the Committee identifies. Distributed authority for the management of a situation is not unusual. For example, whether a radiation hazard is managed under the mandates of the Environmental Protection Agency, the Department of Transportation, or the Nuclear Regulatory Commission will depend on whether the radiation source is in a nuclear-energy producing facility, is in transit, or is being disposed of.

As scientists and engineers, we cannot determine whether FAA alone can address the questions we raise, nor can we easily say whether they are questions of health, safety, or comfort. The legislative branch (if law must be clarified or written) and the executive branch (if, for example, coordination among agencies is required) must sort out responsibilities in appropriate ways.

The Committee has gathered for the first time much important information about a complex environment. As a result of the study, we make one recommendation that clearly will be controversial. It is unanimously and forcefully proposing that smoking be banned on all commercial flights within the United States. The reasons are presented and elaborated in the text and executive summary, but the process by which the decision was reached belongs here.

First, it should be emphasized that the makeup of the Committee was diverse, and only three of the 11 members were physicians with experience in the care of patients crippled or dying as a result of cigarette-smoking. Most of the members are ex-smokers who are admittedly annoyed by cigarette smoke in airliner cabins, as well as other public environments. However, most began the study with the assumption that addicted smokers could not be deprived of their habit over long flights, and therefore smoking could not be prohibited, especially on longer flights. Development of support for a complete ban was gradual, as the evidence of contamination and the impossibility of adequate cleansing of the cabin air became more and more apparent. The coup de grace to smoking in airliners was the realization that diminished ventilation with outside air and increased recirculation of air, a characteristic of almost all new airliner models, will increase previous levels of toxic products of cigarette-smoking in nonsmoking sections of the cabin. When smoking is permitted, the result of these changes places cabin air ventilation in violation of the building codes for most other indoor environments.

We recognize that prohibition of smoking on airplanes will cause discomfort and annoyance among inveterate smokers and the tobacco industry, but it is also likely to be supported by the majority of the flying public and cabin crew members. We hope that the controversies likely to arise regarding this recommendation will not divert deserved attention from the other notable proposals of the report, especially the call for much more research on other aspects of cabin air quality.

To conduct its study, the Committee reviewed the available scientific and technical literature, including characteristics of various models of modern aircraft. It conducted a series of technical meetings and briefings with experts in relevant fields. In addition, members made a number of site visits to evaluate specific aspects of the issues before the Committee. The sites included: National Airport, to examine the cabin air circulation machinery of a TWA MD-80; the FAA Technical Center in New Jersey, to review procedures for testing flammability of cabin materials; the United Airlines flight attendant training Center in Chicago, to gather information about emergency training procedures; and the Boeing Commercial Airplane Company in Seattle, to explore developments in aircraft design. We are grateful to all those who educated and informed us during these visits. The Committee also thanks FAA for its support in supplying the information and assistance we requested. The Committee is unanimous in its praise of National Research Council staff, who worked prodigiously to make our job easier and more effective. Equally important, I thank the Committee members for their hard work in individually reviewing data and writing the text and for their good humor and substantive contributions to our many meetings.

THOMAS C. CHALMERS, CHAIRMAN
COMMITTEE ON AIRLINER CABIN AIR QUALITY

Acknowledgments

The preparation of this report by the Committee on Airliner Cabin Air Quality would not have been possible without assistance from a large number of people and organizations.

We especially wish to thank the Federal Aviation Administration (FAA), the sponsoring agency, for responding to our numerous requests for help in locating and gathering information. In particular, we are grateful to Philip J. Akers, the project officer, and to the following other FAA staff members: Louis C. Bicknese, Charles R. Crane, Edward R. Graves, Andrew F. Horne, Leroy A. Keith, Thomas E. McSweeny, Joseph A. Pontecorvo, William T. Shepherd, Robert N. Thompson, Benjamin H. Tollison, Jr., and the FAA library staff.

Flight attendants have direct experience with airliner cabin environments, and they helped focus attention on issues of air quality, health, and safety. We express our special thanks to Margaret Brennan and Lynne Egge, representing the Joint Council of Flight Attendant Unions, to flight attendants Phyllis W. Conrad, Nancy Garcia, Janna F. Harkrider, Lana Holmes, and Betsy Murtaugh; and to all flight attendants who wrote letters to the Committee.

The Committee visited several facilities to obtain information and observe operations and practices. We wish to thank David J. Shearer, supervisor of airport services for TWA at National Airport, for explaining the ventilation system of an MD-80 aircraft; Constantine P. (Gus) Sarkos and the technical staff at the FAA Technical Center in Atlantic City, for demonstrations of fire testing procedures and full-scale simulation tests; United Airlines personnel Paul Smith, emergency procedures training manager, Janice Northcott, inflight safety manager, and Robert A. McGuffin, regional flight surgeon, for providing information and an opporunity to participate in training exercises; and Boeing Commercial Airplane Company personnel M. E. Kirchner, director of technology, B. C. Hainline, chief engineer, and their colleagues J. N. Bigford, E. E. Campbell, M. T. Katsumoto, G. Veryioglou, and A. S. Yorozu for discussing a variety of topics and demonstrating their planning techniques.

The Committee consulted with a number of experts about various topics. We would like to thank John C. Bailar, William Cain, Frederick B. Clarke, III, Arthur B. DuBois, and Ralph Goldman for their contributions. The Committee gives special thanks to Barry Ryan of Harvard University for developing the mathematical model used in this report.

An open meeting was held to receive comments from the public. We are indebted to Senator Daniel K. Inouye for giving the keynote address. Several people made presentations at the meeting; they are included in the list below.

The Committee thanks all the peer reviewers of the report. Their constructive remarks contributed to the improvement of presentations of technical information and its readability.

So many other individuals and organizations assisted the Committee in various ways that it is difficult to mention them all. Nevertheless, with apologies for whatever inadvertent omissions occur, we shall try:

John F. Banzhaf, III, Action on Smoking and Health

Richard Beckham, American Association for Respiratory Therapy

William M. Beckner, National Council on Radiation Protection and Measurements

Merritt M. Birky, National Transportation Safety Board

Cecil Brownlow, Flight Safety Foundation, Inc.

Phyllis Cleveland, National Aeronautics and Space Administration

J. Donald Collier, Air Transport Association of America

Elliot C. Dick, University of Wisconsin

Pascal Dranitsaris, Ontario Research Foundation
Matthew H. Finucane, Association of Flight Attendants
Hugh B. Fulton, Jr., airline pilot
Sam Giordano, American Association for Respiratory Therapy
Ruth M. Heron, Transport Canada
Rufus R. Hessberg, Aerospace Medical Association
Verena E. Klemm, Independent Union of Flight Attendants
Jane Larkin, American Medical Association
Richard E. Macdonald, Aerospace Industries Association of America, Inc.
Samuel C. Marcus, National Center for Health Statistics
Nora Marshall, National Transportation Safety Board
Robert Millenburg, Smiley, Olson, Gilman and Pangla, Attorneys
Stanley R. Mohler, Wright State University
Philip R. Morey, National Institute for Occupational Safety and Health
Matthew L. Myers, Coalition on Smoking OR Health
Malcolm B. Reddoch, Food and Drug Administration
John P. Reese, Aerospace Industries Association of America, Inc.
James L. Repace, U.S. Environmental Protection Agency
Harold Schaitberger, International Association of Fire Fighters
Allen Schanot, National Center for Atmospheric Research
Sorrell L. Schwartz, Georgetown University Medical Center
Joseph M. Schwind, Air Line Pilots Association
Donald R. Shopland, Office on Smoking and Health
Jim Spahn, National Council on Radiation Protection and Measurements
Martin Speiser, National Transportation Safety Board
Dick Tobiason, National Aeronautics and Space Administration
Jacques Vuille, Transport Canada
Judy A. Weidemeier, The Tobacco Institute
Bertil Werjefelt, Xenex Corporation
Wayne E. Williams, National Transportation Safety Association
P. Wilson, The Guild of Air Pilots and Air Navigators
Edward C. Wood, Flight Safety Foundation, Inc.
Mark Young, The British Air Line Pilots Association.

We would like to express our thanks to the NRC staff for their work in supporting the Committee. Alvin G. Lazen, executive director of the Commission on Life Sciences, provided valuable advice concerning the intricacies of NRC activities. Devra Lee Davis, acting director of the Board on Environmental Studies and Toxicology, gave unstinting attention and support to the study. Edna W. Paulson and the staff of the Toxicology Information Center were of great assistance. Other NRC staff members contributed information and reviewed drafts of various documents for the Committee, among them Stanley M. Barkin, Committee on Toxic Hazards of Materials Used in Rail Transit Vehicles; Henry Borger, Advisory Board on the Built Environment; Stephen L. Brown, Board on Radiation Effects Research; Karen L. Hulebak, Committee on Fire Toxicology; and Diane K. Wagener, Committee on Passive Smoking.

Our thanks to Norman Grossblatt for editing the entire report.

Finally, we wish to express our gratitude to the following NRC staff who directly supported the study. James A. Frazier, the project director, was tenacious and persevering in shepherding us through the study. Andrew M. Pope provided valuable assistance during the formative stages of the study, and Rob Coppock contributed considerably toward the end of our task. Alison Kamat provided invaluable service not only in documenting, locating, and obtaining the large amounts of literature required in the Committee's deliberations, but also in producing the manuscript of the report. We especially thank Judy Tiger for holding us to deadlines, putting together many report drafts, and coordinating a multitude of details.

Contents

Executive Summary

Each year Americans take more than 300 million plane trips, and airliner cabins are the workplace for about 70,000 flight attendants. The health and comfort of these travelers depend on the complex interplay of several factors: the adequacy of ventilation systems affecting the amount of cigarette smoke, microorganisms, and other contaminants in the cabin air; the use of fire-retardant materials in cabin equipment and furnishings; the availability and ease of use of breathing and other emergency equipment; and the clarity of special and emergency instructions.

Although such devices and procedures are usually taken for granted, Congressional hearings during 1983–1984 revealed that information on airliner cabin air quality was contradictory. Flight attendants and others testified about inadequate ventilation and other problems with the cabin environment that caused discomfort. Representatives from the airline industry and federal regulatory agencies argued that present standards for the airliner cabin were adequate to protect the health and safety of travelers.

Under Public Law 98-466, Congress stipulated that the National Academy of Sciences enter into a contract with the Federal Aviation Administration (FAA). The Academy was asked to determine whether such aspects of cabin air as the quantity of outside air, the quality of onboard air, the extent of pressurization, the characteristics of humidification, the presence of cosmic radiation, contaminants (such as bacteria, fungi, and other microorganisms), and pollutants (such as environmental tobacco smoke, carbon monoxide, carbon dioxide, and ozone) could be responsible for health problems in the long or short run; to recommend remedies for problems discovered; and to outline the safety precautions necessary to protect passengers in event of in-flight fires, which produce smoke and fumes. Accordingly, the Committee on Airliner Cabin Air Quality was established in the National Research Council's Commission on Life Sciences. This report summarizes the findings of the Committee's 18-month study of relevant issues. The investigation covered five general subjects:

- Cabin air quality: including potential health effects of reduced ventilation and of contamination by chemicals, microorganisms, other allergens, tobacco smoke, and ozone.
- Cabin environment: health effects of reduced pressure and of cosmic radiation.
- Emergency procedures: control of fires and toxic fumes, use of emergency breathing equipment, and adequacy of emergency instruction given passengers.
- Regulations: regulations established by U.S. and foreign agencies.
- Records: status and adequacy of medical statistics on air travel, of records on airline maintenance, and of records on operating procedures.

The Committee relied heavily on published material—articles in scientific and medical journals and government and industry publications. FAA provided accident data and information on continuing investigations. Members of the Committee also visited government, airline, and industry groups to review fire testing, crew training facilities, and research programs on cabin ventilation. Relevant comments and information were received from the general public and other interested groups at an open hearing and were reviewed by the Committee.

In formulating its conclusions and recommendations, the Committee attempted, but abandoned, the separation of issues of health from those of safety. However, under current statutes and administrative orders, no federal office has direct responsibility for health effects associated with air travel. This lack of correspondence between the issues as conceived by the Committee and the responsibilities of federal agencies contributed to the difficulty of the Committee's work. The Committee believes that the health effects associated with air travel should be within the purview of a federal agency.

CABIN AIR QUALITY

In assessing the overall quality of onboard air, the Committee determined the range of outside-air ventilation rates on the U.S. fleet by reviewing manufacturers' design specifications, airline load-factor data, and operating procedures. No data were available on actual measured airflow in the fleet.

The Committee found that, if the lowest rate of ventilation permitted by current equipment design were used under conditions of full or nearly full passenger loads, the resulting ventilation rate would be at the minimum determined to provide acceptable air quality when smoking is not permitted and other contaminant sources are not present . In the absence of sources of contamination, this rate does not constitute a health hazard.

In particular, the Committee noted that the flow rate of outside air varied from below 7 cubic feet per minute (cfm) per economy-class passenger to 50 cfm per first-class passenger. Cockpit ventilation rates are often as high as 150 cfm per crew member; this higher rate, however, is provided to meet avionic and electronic equipment cooling loads, rather than for reduction of contaminant concentrations. These rates compare with a ventilation rate of 5–7 cfm/person established for other types of vehicular travel that have nonsmoking sections, including passenger and commuter trains and subways. It should be noted, however, that these other ventilation standards do not consider possible synergistic effects of the low relative humidity encountered in aircraft.

Another important consideration is the adequacy of oxygen supply—because the normal requirement of air to meet oxygen needs for sedentary people is only 0.24 cfm/person, the amount of oxygen is sufficient in aircraft even at the lowest rate of flow of outside air.

Nevertheless, a minimal ventilation rate for airplane passenger cabins is not defined under FAA regulations, which specify ventilation rates only for flight crew compartments. Actual cabin airflow is seldom measured once an aircraft is in service; and flow can be reduced by deterioration in equipment performance. A data collection program that measures airflow and contamination in airplane cabins should be implemented.

Carbon Dioxide

The Committee's efforts in evaluating contaminant concentrations were hampered by an almost complete absence of reliable data. The carbon dioxide concentration associated with a given ventilation rate, however, can be estimated with confidence. For a rate of 9.7 cfm/occupant, the carbon dioxide concentration would be about 0.15%, or 1,500 ppm. No adverse health effects of carbon dioxide would be noted at this concentration, but the FAA standard for aircraft allows for 20 times this concentration. This is considerably higher than standard concentrations permitted by the Occupational Safety and Health Administration (OSHA) and the American Society of Heating, Refrigerating, and Air-Conditioning Engineers (ASHRAE) for other types of indoor environments. The FAA standard is much higher than standards for other confined environments. The Committee recommends that FAA review its carbon dioxide standard.

Humidity

In addition to carbon dioxide, relative humidity in the cabin at flight altitude is predictable, depending only on cabin ventilation rate, passenger load factor, temperature, and pressure. With a range of standard cabin ventilation rates, the relative humidity varied from 23% to less than 2%. After 3 or 4 hours of exposure to relative humidity in the 5–10% range, some passengers experience discomfort, such as dryness of the eyes, nose, and throat. However, the Committee could find no conclusive evidence of extensive or serious adverse health effects of low relative humidity on the flying population that would justify recommending a regulation to add supplementary humidification systems to aircraft.

Ozone

Ozone has been measured at concentrations above 0.8 part per million by volume (ppmv) in the cabin during flight above the tropopause and during periods in which there is increased vertical air-exchange between the stratosphere and the troposhere. This relatively high concentration can be reduced if ozone control equipment has been installed and is operating or if altitude and route limitations are imposed. In comparison with the observed ozone concentration of 0.8 ppmv, compliance with existing standards would limit ozone concentration to a maximum of 0.25 ppmv at equivalent sea level pressure. Standards also limit the time-weighted average ozone concentration for any flight segment of over 4 hours to 0.1 ppmv.

The Committee could find no documentation of the effectiveness of the various methods being used by the airlines to control ozone. Therefore, the Committee suggests that FAA carry out a carefully designed program to ensure that cabin ozone concentrations comply with Department of Transportation regulations.

Environmental Tobacco Smoke

A contaminant in aircraft cabins that can be detected by its characteristic odor and visibility is environmental tobacco smoke (ETS)—the combination of exhaled mainstream smoke and the smoke generated by smoldering cigarettes. ETS is a hazardous substance and is the most frequent source of complaint about aircraft air quality. In the past, ventilation systems on aircraft were designed to control odor and irritation from cigarette smoke on the assumption that smokers are randomly distributed throughout the aircraft. However, separation of smokers and nonsmokers into separate zones is now federally mandated. Because of the high concentration of ETS generated in the smoking zone, it cannot be compensated for by increased ventilation in that zone. Moreover, strict separation of the airplane into smoking and nonsmoking zones does not prevent exposure of flight attendants and nonsmoking passengers to ETS, because of the location of galleys and lavatories in the smoking areas. Smoke exposure can become significant in aircraft with outside-air flow rates as low as 7 cfm/passenger. Even a ventilation (airflow) rate of 14–15 cfm/passenger consists of as much as 50% recirculated, and possibly smoky, cabin air.

It is not known how often operating procedures are used that can decrease actual ventilation rates and increase contaminant concentrations. The Committee found no published peer-reviewed data on ETS concentrations in cabins. Although the adverse effects of ETS are still under investigation, the Committee feels that this potential threat to the health of nonsmoking passengers and flight attendants should not be ignored, especially because flight attendants on some airlines can fly up to the twenty-eighth week of pregnancy. It is highly probable that eye, nose, and throat irritation will increase among airline passengers as outside-air ventilation rates are decreased and recirculation is increased to improve fuel efficiency.

The Committee considered several ways of reducing ETS concentrations in aircraft. Any solution requiring structural or engineering changes—such as markedly increasing ventilation, moving lavatories and galleys, and separating smoking compartments with physical barriers—appears economically infeasible. Increasing ventilation of the smoking zone to the point where it is in compliance with ASHRAE guidelines and eliminating recirculation on existing aircraft does not appear technically feasible. The amount of air that would be required could exceed the engine bleed capacity and in all cases would reduce the range of the aircraft, the payload, or both. Injection of large volumes of air into the cabin would create unacceptable air velocities and result in passenger discomfort. In contrast, the Committee feels that a return to the random distribution of smokers throughout the cabin to reduce overall ETS concentration would be unacceptable to a majority of the traveling public.

Cigarette-smoking has been implicated in a small number of in-flight fires, and thus presents a potential threat to safety.

The Committee recommends a ban on smoking on all domestic commercial flights, for four major reasons: to lessen irritation and discomfort to passengers and crew, to reduce potential health hazards to cabin crew associated with ETS, to eliminate the possibility of fires caused by cigarettes, and to bring the cabin air quality into line with established standards for other closed environments.

Aerosols

Evaluation of the degree of health hazard associated with exposure to biologic aerosols was impossible, because of the lack of data on their concentrations in aircraft cabins. There is an urgent need for studies of potentially infectious airborne agents under routine flight conditions. In the meantime, the Committee's recommendations regarding control of infection through ventilation must be based on similar occupancies (trains and subway cars) for which ventilation standards have been established.

Because a likelihood of occurrence of epidemic disease when forced-air ventilation is not available on the ground has been demonstrated, the Committee recommends that a regulation be established that requires removal of passengers from an airplane within 30 minutes or less after a ventilation failure or shutdown on the ground and maintenance of full ventilation whenever onboard or ground air-conditioning is available.

The Committee also recommends that maximal airflow be used with full passenger complements to decrease the potential for microbial exposure and that recirculated air be filtered (to remove particles larger than 2–3 μm) to reduce microbial aerosol concentrations.

The Committee found no studies of the concentrations of other contaminants—such as volatile organic compounds or substances that might be emitted from disinfectants or cleaning materials—and therefore cannot assess their potential health hazard to passengers or crew members.

Because the Committee found only sparse data on air quality and contaminants in aircraft, it undertook to have a multizone computer model of an aircraft ventilation system developed for its use in calculating contaminant, water vapor, and carbon dioxide concentrations in various cabin zones. The model was used to calculate average and peak concentrations of contaminants in smoking and nonsmoking zones. The effects of reduced flow, recirculation, and filter efficiency were analyzed. Cabin smoke from various onboard cabin fire scenarios can be evaluated with models of this type to develop optimal procedures for control of smoke under emergency conditions. This model is available to FAA.

Cabin Environment

Two unrelated factors of the cabin environment affect airline passengers: pressure and cosmic radiation.

Pressurization of the cabin to equivalent altitudes of up to 8,000 ft, as well as changes in the normal rates of pressure during climb and descent, might pose a risk to or create discomfort for some segments of the population. At an altitude of 8,000 ft (or above if a mistake were made), people with cardiopulmonary disease might be at some risk. Persons suffering from upper respiratory or sinus infections, children, and infants might experience some discomfort or pain because of pressure changes during climb and descent. Injury to the middle ear can occur in susceptible people, but is rare.

Other groups that could be at various degrees of risk include those with chronic pulmonary problems, anemia or sickle-cell disease, gastrointestinal problems, neuropsychiatric symptoms, or recent abdominal or eye surgery. Pregnant women should not fly beyond 240 days; pregnant women with a history of spontaneous abortions should not fly; and scuba divers should not fly sooner than about 12–24 hours after diving.

The Committee concluded that current pressurization criteria and regulations are generally adequate to protect the traveling public. However, the medical profession should use a more efficient system to warn those with existing medical conditions who are more susceptible to changes in pressure or to long exposure to low-pressure that there might be some hazard to their health.

Although the dose-equivalent rate of cosmic radiation is significantly higher at airplane cruise altitudes and above than at ground level, cosmic radiation associated with subsonic commercial flights does not pose a serious health risk to the general public. However, it is likely that some flight and cabin crew members will receive 100–200 mrems/yr. That is below the 500-mrem/yr recommended maximum for any member of the general public. Inasmuch as radiation exposures are additive and assumed to be linear, the additional radiation received during high-altitude flying should be considered in the estimates of total dose, which includes the radiation that might be received as a result of living at high altitude or from medical or dental x rays.

This report draws attention to the potential hazard to full-time flight attendants flying high-latitude routes, who might be exposed to cosmic radiation equivalent to radiation from thoracic or abdominal medical x rays. Such medical x rays are to be avoided during pregnancy. FAA should consider rule-making that restricts exposure of pregnant flight crew and cabin crew members. In addition, FAA should investigate total radiation exposure of flight crew and cabin crew members through the use of a statistical sample of full-time employees and should require airlines to provide precautionary information to their flight attendants about radiation exposure.

EMERGENCY SITUATIONS AND PROCEDURES

The Committee reviewed emergency procedures and cabin crew training for evacuation of the cabin in emergencies or after survivable crashes and the procedures for use after cabin depressurization. Several members of the Committee participated in an emergency evacuation exercise. The Committee also investigated fire test procedures for cabin materials, firefighting techniques, and emergency breathing equipment for cabin crews.

As any air traveler can observe, many passengers ignore or pay little attention to passenger safety briefings, in spite of the fact that retention of the information presented can mean the difference between survival and death in emergency situations. The Committee approves of current efforts to base passenger safety briefings and written materials on empirical testing of comprehension and retention. However, the Committee believes that it is also important to understand how passengers recall that information and respond under the stress of emergencies. The Committee suggests that FAA or appropriate industry organizations consider the advisability of developing an empirical research program to examine passenger response to safety instructions under routine and emergency conditions and revise them as appropriate . Consideration should be given to running some quizzes during a flight to see, for example, what proportion of passengers have retained the key features of the safety briefing.

The Committee recommends that FAA require that information on proper response to fire emergencies be included in oral and written passenger safety information.

In general, the FAA program on flammability testing is excellent, and its research efforts to improve testing methods are appropriate and valuable. The recently issued FAA flammability standards for seat cushions and cargo compartment liners will reduce in-flight and postcrash fire hazards. The Committee feels that continuing research is also needed in materials development. Although FAA standards are met by currently available materials, other materials exist that, with further development, would far exceed current standards and would provide substantially increased fire protection in aircraft.

The Committee noted that current emergency procedures for smoke removal recommend that the cabin be depressurized to 10,000 ft. This procedure is ineffective and should be discontinued.

FAA recently proposed standards that would require that protective breathing devices be available to airliner crew members for firefighting. One such device is to be stored within 3 ft of each required fire extinguisher. However, there are generally more crew members than fire extinguishers, and the Committee recommends that FAA review the proposed rule on protective breathing devices for crew members to ascertain the desirability of supplying such equipment for all crew members, rather than limiting it to the persons expected to be involved in firefighting. In addition, the Committee suggests further evaluation of the potential of emergency breathing equipment for all cabin crew members to improve safe and expeditious evacuation of passengers in fire emergencies.

A rule requiring protective breathing devices for passengers was proposed by FAA in 1969, but later withdrawn. These devices have since been further developed and evaluated. The Committee recommends that FAA re-examine passenger protective breathing devices and consider requiring that such equipment be available in case of in-flight and postcrash fires.

WORLDWIDE AIRLINE REGULATIONS

The Committee was charged with performing a comparison of foreign industry practices, regulations, and standards, and has gathered relevant information applicable to the issues addressed in this study. Although some differences from those in the United States have been noted, they do not appear to be significant. The Committee feels that greater effort along these lines is not warranted.

FEASIBILITY OF DATA COLLECTION

Empirical evidence is lacking in quality and quantity for a scientific evaluation of the quality of airliner cabin air or of the probable health effects of short or long exposure to it. Standards directly applicable to commercial aircraft have not been established for cabin ventilation rates, environmental conditions, and air contaminants, and adequate data on these factors are not available. The Committee therefore recommends that FAA establish a program for the systematic measurement, by unbiased independent groups, of the concentrations of carbon monoxide, respirable suspended particles, microbial aerosols, and ozone and the measurement of actual ventilation rates, cabin pressures, and cosmic radiation on a representative sample of routine commercial flights. These findings should be subjected to peer review . This would provide a basis for establishing appropriate standards if justified and for requiring regular monitoring if necessary.

The Committee recognizes the extreme difficulty of interpreting data on the health effects of air travel, but believes that several kinds of data can be collected. The Committee recommends that FAA establish a program to monitor selected health effects on airliner crews.

Air carriers are required to report to FAA all uses of the recently mandated medical kits during the first 24 months. The Committee recommends that FAA collect these data in such a way as to permit comparison of onboard incidents with those in other settings.

Introduction

Air travel has become an essential form of transportation in modern society. It has also been one of the safest, even though aviation accidents have received considerable attention in recent years and have caused public concern about personal safety. Concern for the quality of air in the passenger cabins of commercial airliners has been publicized, but it has focused on occupational exposures of cabin crews. Rising fuel costs might have prompted airlines to reduce the amount of outside air in the ventilation of passenger cabins to conserve fuel, consequently adversely affecting the air quality. Finally, new models of aircraft use recirculation of cabin air to a greater degree than older models in the fleet, so the general tendency is toward the use of less outside air.

The specific concerns regarding the quality of cabin air include not only the amount of outside air, but also the adverse effects that might result from exposure to this unique confined environment. In aircraft, people are exposed to a particular combination of low relative humidity, reduced air pressure, presence of ozone and other pollutants (some of which have been demonstrated to be harmful to human health), and increased cosmic radiation.

In the last year, 28% of the general public took at least one trip by air, and 5% of those who flew took 10 or more trips. In addition, more than 40,000 flight attendants are exposed to the cabin environment for an average of approximately 900 h each year. Yet, in the face of the knowledge of these acute and chronic exposures to pollutants with proven health effects, very little research has been done to characterize either the quality of the air in airliner cabins or the potential health effects of exposure to that environment.

Many airline travelers complain about cabin air quality. The nature of and reasons for their complaints are important clues to the problem. Complaints from airline passengers about catching colds or experiencing other health problems as a result of air travel are not uncommon. Although airliner cabins are divided into smoking and nonsmoking sections, many travelers still insist on being seated as far as possible from the smoking section; complaints by some groups have led to the suggestion of eliminating smoking in aircraft. Concern has also been registered about the possible relation of this environment to acute exacerbations of underlying chronic diseases, such as allergic rhinitis or asthmatic attacks, and about the adequacy of onboard medical equipment and the availability on every flight of trained personnel to handle emergency situations.

For years, flight attendants have reported various health problems—from chronic bronchitis to difficulties in pregnancy—that they have attributed to their occupational exposures. Flight attendants' careers have become longer, and female flight attendants are permitted to work until late in pregnancy. Furthermore, a larger portion of the general public, some with health conditions that might make them more susceptible to the airliner cabin environment, are now flying. It is therefore important to understand the potential for adverse health effects of chronic exposure to airliner cabin air.

Onboard fires are a special condition of cabin air quality that is the basis of additional concern because they produce large quantities of smoke and toxic fumes. There is concern about the adequacy of emergency fire procedures and equipment, including both firefighting and protective breathing devices.

The scientific community and more recently the general public have become aware of the effects of pollution in confined spaces. Unlike other modes of transportation or other public spaces, airliners do not offer users the freedom to open a window, move away, or step outside if the cabin air is not suitable.

Throughout the course of its work, the Committee on Airliner Cabin Air Quality had to confront the problem of answering questions on which almost no scientifically valid data were available. The study began with a public meeting, whose purpose was to collect information for the Committee to review. In addition, several experts made presentations to the Committee on various relevant issues throughout its work. The Committee also inspected the environmental control system of an MD-80 airplane and visited research facilities—such as the Federal Aviation Administration (FAA) Technical Center, the United Airlines flight attendant training center, and the Boeing Commercial Airplane Company—to gain a better understanding of the state of the science related to the issues at hand. Information was obtained from the airline industry and flight attendants' unions, and computerized FAA data banks were used for accident and incident data.

STRUCTURE OF REPORT

This report is the product of extensive Committee deliberations on the issues associated with the potential health effects of exposure to airliner cabin air. Chapter 1 describes the magnitude of the population exposed to cabin air. Chapter 2 discusses the current environmental control systems on commercial passenger aircraft, and Chapter 3 describes airliner safety procedures, equipment, and passenger instructions. Chapter 4 discusses the effects of cabin fires and depressurization on air quality. Chapter 5 identifies the sources and exposures of cabin air pollutants, and Chapter 6 discusses the reported health effects associated with cabin air. Chapter 7 considers the desirability and feasibility of collecting additional data.

PROBLEMS IN STUDYING AIRLINER CABIN AIR QUALITY

The Committee faced two fundamental and related problems in its attempt to assess exposure to pollutants in airliner cabins: although their presence is known or suspected, very few data are available on the concentrations of pollutants of interest; and National Research Council committees do not generally conduct basic research and gather their own primary data, but rather rely on the available, published, peer-reviewed literature. This Committee explored the idea of collecting primary data, but was unable to complete a sampling program within its schedule and budget constraints. The Committee therefore identified and reviewed other relevant studies and models of indoor air pollution that could be extrapolated to the airliner environment and, with the assistance of consultants from Harvard University, developed a computer model of pollution in the airliner cabin to simulate typical exposures under various operating conditions.

Although most of the concerns raised about airliner air quality have been related to the cabin, the only existing aircraft regulations that specify ventilation rates apply to the flight deck (cockpit), not to the cabin. That is the case because of safety considerations, which dictate that the cockpit be adequately ventilated —both to provide a safe working environment for pilots and to cool sensitive equipment. Air-exchange rates in the cockpit are typically more than 10 times those in the passenger compartment. The Committee chose not to address the issue of cockpit air quality specifically, however, because the conditions and issues are different. The Committee focused its attention on cabin air quality and chose not to expand the scope of its study to include cockpit conditions.

Airliner cabin air consists primarily of air drawn from compressors in the engine (bleed air) and often contains recirculated air from within the cabin. On the ground, some aircraft with vapor-cycle cooling can use primarily recirculated air. There are no federal standards regulating ventilation, relative humidity, and mixing efficiency, all of which greatly affect the quantity, distribution, and overall quality of air in the cabin. Instead, individual manufacturers set performance requirements for airliner environmental control systems; as a result, design and performance can vary among different models of aircraft. If a system fails, emergency "standard operating procedures"—some set by government regulation and some by individual airlines—govern the operation of backup systems and in-flight procedures for ventilation and air quality. Given human nature, however, and the fact that these procedures are often carried out under stressful conditions (i.e., when normal systems have failed), there might be variation in the actual performance of these procedures and operation of emergency equipment.

Moreover, experts disagree as to whether aircraft cabins are adequately ventilated even during normal operation.

The Committee examined standards developed by federal agencies and other organizations for relevant pollutants in other environments, both indoor and outdoor, ranging from offices and homes to spaceships and submarines—the last two of which share some properties with the airliner cabin environment (high-density occupancy in confined space, air recirculation, and problems with humidity)—and compared these standards with the concentrations commonly found in aircraft cabins (see Table I-1). The documents developed in support of federal standards for other environments also provided useful information for making inferences about the applicability of standards to the aircraft cabin. However, those standards are not directly applicable to airliner cabins.

The three substances in aircraft cabin air that FAA regulates are ozone, carbon monoxide, and carbon dioxide. Ozone may not exceed 0.25 ppm at any time and may not exceed 0.1 ppm for periods longer than 3 h.[3] Carbon monoxide in excess of 1 part in 20,000 parts of air (50 ppm) is considered hazardous.[7] Carbon dioxide in excess of 3% by volume (sea level equivalent), or 30,000 ppm, is considered hazardous in the case of crew members,[7] but may be allowed in crew compartments if appropriate protective breathing equipment is available.

As to relative humidity and low-pressure, the Committee relied on reports from the toxicologic, clinical, and epidemiologic literature and estimated health effects associated with combinations of humidity, pressurization, and pollutant exposures.

The Committee could find no published data on biologic contamination in aircraft cabins, although instruments are available for measuring many biologic contaminants and measurements have been made in other confined spaces. Therefore, the Committee felt that it would be especially worth while to determine the feasibility of detecting and measuring these contaminants.

TABLE I-1 U.S. Standards for Exposure to Selected Substances

Substance	EPA Ambient Air Quality Standards[4]	OSHA Occupational Health Guidelines[5]	ACGIH Threshold Limit Values (Time-Weighted)[1]	ACGIH Short-Term Exposure Limits Standards[1]	ASHRAE Indoor Air Quality[2]	14 CFR 25.831[7]
Carbon dioxide	None	5,000 ppm	5,000 ppm	15,000 ppm	2,500 ppm	3%
Carbon monoxide	9 ppm (8 h) 35 ppm (1 h)	50 ppm	50 ppm	400 ppm	9 ppm (8 h)	50 ppm
Nitrogen dioxide	0.053 ppm (annual)	5 ppm	3 ppm	5 ppm	0.05 ppm (annual)	None
Particles (TSP)	75 µg/m^3 (annual) 260 µg/m^3 (24 h)	Nuisance particles (respirable dust): 15 mg/m^3 (8 h)	Nuisance particles: 5 mg/m^3	None	75 µg/m^3 (annual) 260 µg/m^3 (24 h)	None
Ozone	0.12 ppm (1 h)	0.1 ppm	0.1 ppm	0.3 ppm	0.05 ppm	0.25 ppm[a] 0.1 ppm (3 h)
Cosmic radiation	None	None	None	None	None	None

[a] Above flight level 320 (32,000 ft at standard atmosphere).

The Committee also considered the general issue of exposure. Exposure to co smic radiation is difficult to predict, because it depends on the variable solar flux and the amount of radiation reaching our atmosphere. Clearly, an occupant's location in the aircraft affects his or her exposure to air contaminants that have point sources, such as cigarette-smoking, food-based materials, specially applied cleaning agents, and specific infectious agents carried by passengers. Other possible pollutants, such as ozone and hydrocarbons, might be distributed more uniformly. The most commonly noticeable differences in passenger air quality are associated with locations in or near designated smoking zones and with the density of such smoking zones in a given area. Because of aircraft cabin airflow patterns, there can be significant differences in the exposure of passengers seated toward the tail, in the middle, near the lavatories and galleys, and in the forward compartments of large aircraft. Flight galleys and lavatories, which have local vents and fans, have patterns of ventilation and sources of air contamination different from those in the rest of the passenger compartments.

There are differences in exposures to pollutants during boarding (e.g., in the waiting areas) and during the various stages of flight. During boarding—with cabin doors open, no smoking allowed, and little lavatory use or food preparation—air quality is different from that during flight. The increase in smoking frequency after the "no smoking" sign is turned off leads to a worsening of air quality in the smoking area and contiguous zones. Extended periods of holding at the dock, taxiing to the runway, or awaiting clearance to take off or land can adversely affect cabin air quality. Unusual events can also have dramatic effects on cabin air quality—e.g., mechanical problems; sudden changes in heating, ventilation, and air-conditioning controls (HVAC) or in cabin pressure; spills or fluid leakage; and undetected, smoldering electric or cigarette fires.

The presence of so many complex variables that affect cabin air quality led the Committee to commission the preparation of a mathematical model that could be used to calculate concentrations of substances in the air of different parts of the cabin with different passenger load factors and operating modes. The model described in Appendix A is based on sound physical principles, but has not yet been verified in practice.

PROBLEMS IN STUDYING THE HEALTH EFFECTS ASSOCIATED WITH AIRLINER CABIN AIR QUALITY

Data relevant to the health effects of airliner cabin air quality can be considered in several ways, for example, by types of persons exposed, by acute and chronic health outcomes of concern, and by source and type of data. Each implies tradeoffs.

Health effects, even those of great concern, can be hard to detect, measure, and attribute to specific causes (such as a component of cabin air). The reasons, which are numerous, include the lack of baseline observations on most persons who fly and who have been adversely affected, ethical constraints on and practical infeasibility of followup of persons who fly and might be adversely affected by the various features of cabin air, the imprecise nature of many relevant symptoms of acute effects (such as tightness of the chest), the rarity of chronic health problems directly attributable to cabin air, and the self-selection characteristics of both crew and passengers. These difficulties are mitigated in some types of research studies on biologic correlates of illness, for example, respiratory function studies of crew members before, during, and after flight. However, although acute changes in FEV_1 (forced expiratory volume—the maximal amount of air that can be expelled in 1 s) are well correlated with acute respiratory symptoms and disease, they are subject to a potentially misleading measurement bias. It is not at all clear that a small decrease in FEV_1 in the average healthy person has any health (or regulatory) meaning. Thus, there is a three-way tradeoff that involves feasibility of detecting effects through epidemiologic studies, precision of measurement, and relevance to public health.

Several problems complicate assessments of the effects of cabin air quality on the health of passengers and crew. Biochemical measurements, such as those of the absorption and excretion of toxic products, can be made with precision and reliability, but the responses of people exposed to those substances are most difficult to ascertain and to assess. The prevalence and incidence of abnormal symptoms among exposed people can be determined through questionnaires and direct questioning, but questions in both approaches must be carefully worded to avoid suggesting answers and to avoid unintentional inherently biased responses. Some questions not apparently related to the effects of the substances at issue must be included, and people in control groups must be selected with care and treated in exactly the same way as the people in the test groups. When selecting controls, one must keep in mind the likely differences between the types of people who might apply for jobs in the air and jobs on the ground and the possible impacts of those differences on the end points being measured. Biases resulting from the expectation of compensation for occupation-related illness must also be considered. None of the studies of health effects found by the Committee satisfied these criteria of reliability.

The Committee discovered that, with the exception of pilots, few routine health data are collected by the airlines on either the flying public or airline personnel. In the case of the public, only acute episodes that occur on planes are noted. As to chronic effects associated with intermittent or continuous small exposures, no routine monitoring is available. The health of pilots has been studied several times from the standpoint of the safety of the other occupants of the plane, who depend on them, but little information is available on the impact of flying on the pilots' health. Even less is available on the health of cabin attendants.

The scientific process involves the collection of reproducible facts, and scientific evidence is considered valid only if it is reproducible under similar circumstances or if lack of reproducibility can be explained. In medical surveys, findings tend not to be reproducible when the observer's subjective error is large and thus permits ample opportunity for erroneous or distorted conclusions to be drawn. Few researchers check for systematic internal errors; almost without exception, the authors of the studies reviewed for this report did not measure observer error either by duplicating observations or by attempting to control observer bias through the use of double-blind procedures, regardless of whether the tests were laboratory or clinical examinations.

Demonstrating that a specific health effect is due to a specific cause is usually difficult. There might be limited evidence that exposure to particular characteristics of aircraft cabin air causes acute symptoms or measurable changes in physiologic function, but even this relatively easily found evidence has not been systematically recorded or assembled. Furthermore, very few studies of air pollutants under normal flight conditions and of the reactions of airline personnel to the pollutants have been carried out, and none qualify as exemplary scientific efforts.

Some health-related topics were considered by the Committee to be outside its province to investigate, e.g., the availability of medical kits and training of flight attendants in emergency medical procedures. However, the Committee noted that the recent regulation that such kits be used[6] contains a requirement that all medical emergencies resulting in the use of such kits be reported annually during the first 2 years after implementation of the regulation. This reporting will provide a unique aid to epidemiologic surveys of events possibly related to cabin air quality (see Chapter 7). The seemingly unrelated problem of evacuation time and instructions in the case of fire was considered by the Committee to be within its charge, because death from smoke inhalation occurs almost every year in airplanes and is a most poignant example of the effects of airliner cabin air pollution (see Chapter 4).

REFERENCES

1. American Conference of Governmental Industrial Hygienists. In Documentation of the Threshold Limit Values. 4th ed. Cincinnati, Ohio: American Conference of Governmental Industrial Hygienists, 1984.

2. American Society of Heating, Refrigerating, and Air-Conditioning Engineers, Inc. ASHRAE Standard: Ventilation for Acceptable Air Quality. ASHRAE 62-1981. Atlanta, Ga.: American Society of Heating, Refrigerating, and Air-Conditioning Engineers, Inc., 1981.
3. Cabin ozone concentration. Code of Federal Regulations, Title 14, Pt. 25.832. Washington, D.C.: U.S. Government Printing Office, 1985.
4. National primary and secondary ambient air quality standards. Code of Federal Regulations, Title 40, Pt. 50. Washington, D.C.: U.S. Government Printing Office, 1985.
5. Toxic and hazardous substances. Code of Federal Regulations, Title 29, Pt. 1910.1000. Washington, D.C.: U.S. Government Printing Office, 1985.
6. U.S. Federal Aviation Administration. Emergency medical equipment: Final rule. Federal Register 51(9 Jan.):1218–1223, 1986.
7. Ventilation. Code of Federal Regulations, Title 14, Pt. 25.831. Washington, D.C.: U.S. Government Printing Office, 1985.

1

Profile of Commercial Air Travel

This chapter provides background information on aspects of commercial air travel pertinent to the study of airliner cabin air quality. The study of cabin air quality must take into account both routine flights and emergency situations. Therefore, this chapter presents statistics regarding routine travel—including data on passengers, flight and cabin crews, and aircraft—and enumerates the major types of emergencies that affect air quality, including smoke, fumes, fire, explosion, and depressurization.

PASSENGERS

For most Americans, exposure to an airliner cabin environment is infrequent and brief, although 70% of American adults have flown at least once. According to a Gallup survey conducted in the summer of 1985, 28% of American adults had flown in the preceding 12 months; 14% had taken only one trip, and 1%, 10 or more.[6] Of those who flew in the 12 months before the survey, 50% had taken only one trip and 5%, 10 or more (see Table 1-1). Passenger demographic data appear in Table 1-2.

In 1984, 343,264,000 passenger enplanements occurred on U.S. scheduled airlines, for a total of 304,458,727,000 revenue passenger miles. [2] Figure 1-1 illustrates passenger enplanements projected through 1996; Figure 1-2 shows historical and projected passenger load factors. Table 1-3 lists numbers of passengers carried and total revenue miles by selected airlines in 1984.

TABLE 1-1 Frequency of Flying Among the General Public, 1985a

No. Trips in Preceding 12 Months	% of Adult Population	% of Airline Passengers
1	14	49.8
2	7	22.8
3	2	8.0
4–6	3	11.4
7–9	1	3.1
10+	1	4.9
Total	28	100.0

[a] Data from Gallup.[6]

TABLE 1-2 Demographic Characteristics of Airline Passengers, 1984a

Descriptor	Proportion of Passengers, %	Proportion of U.S. Population, %[b]
Age group:		
<18	11	27
18–24	10	13
25–34	21	17
35–44	23	13
45–54	16	9
55–64	10	9
65+	9	12
Sex:		
Male	54	49
Female	46	51

[a] Data from U.S. Travel Data Center[15] and U.S. Bureau of the Census.[10]

[b] Based on 1983 estimates.

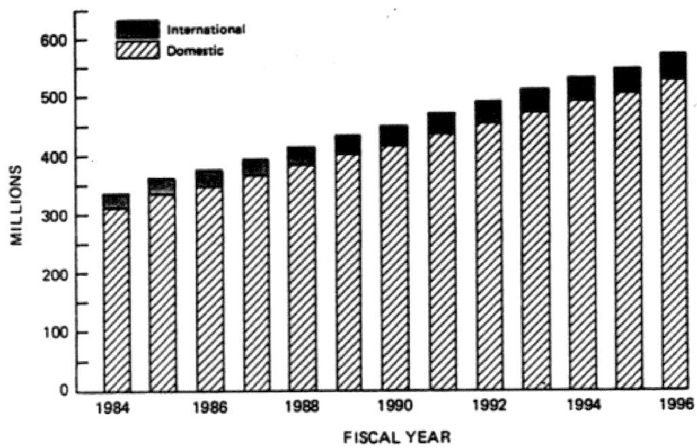

FIGURE 1-1
Scheduled passenger enplanements on U.S. certificated air carriers. Reprinted from U.S. Federal Aviation Administration.[11]

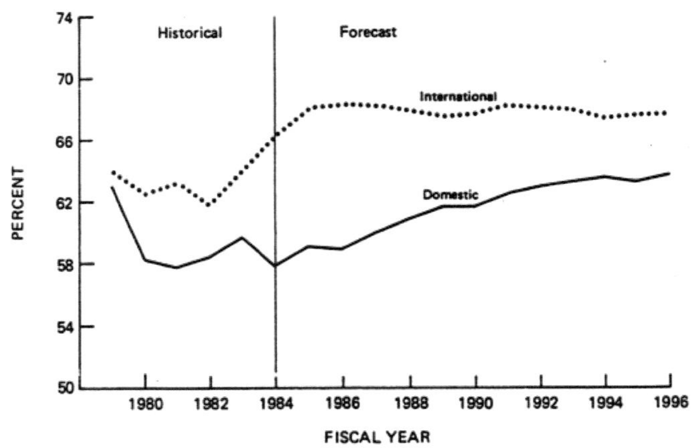

FIGURE 1-2
Passenger load factor on U.S. certificated air carriers. Reprinted from U.S. Federal Aviation Administration.[11]

TABLE 1-3 Airline Passengers and Passenger Miles Flown, 1984a

Airline	Passengers, thousands	Revenue Passenger Miles, thousands
United	41,010	46,037,064
Eastern	38,081	29,359,288
Delta	37,341	27,040,102
American	34,123	36,702,296
Trans World	18,487	28,296,956
US Air	17,047	8,190,589
Republic	15,177	8,509,948
Piedmont	14,274	6,227,641
Pan American	13,913	28,066,826
Northwest	13,216	19,772,356
Southwest	12,052	4,669,435
People Express	11,775	7,770,945
Continental	11,115	10,923,395
Western	10,638	9,396,580
Pacific Southwest	7,830	3,047,338
Frontier	7,048	4,464,168
Ozark	4,949	2,693,866
Air Cal	3,990	1,548,506
Hawaiian	3,022	403,857
New York Air	2,793	937,102
Alaska	2,543	1,841,212
American West	2,398	1,247,134
Aloha	2,346	392,421
Braniff	2,176	1,885,619
Muse	1,980	925,083
Northeastern	1,655	1,652,119
Midway	1,464	747,428
World	1,410	3,889,001

[a] Data from Air Transport Association of America.[2]

FLIGHT ATTENDANTS AND FLIGHT CREW

The populations most exposed to the aircraft environment are flight attendants and flight crew members. These two groups, however, are exposed to different conditions. Flight attendants spend almost all their time in the passenger cabin and galleys, and the flight crew spend almost all their time in the cockpit. The cockpit and passenger cabin are ventilated separately, and the former has a much higher air-exchange rate and hence generally cleaner air.

The Federal Aviation Administration (FAA) restricts most pilots to a maximum of 100 h of flight time per month,[5] and labor contracts impose an even lower limit. The flight time of cabin crew is not regulated, although the Association of Flight Attendants (AFA) has petitioned FAA for rule-making to establish maximal duty hours and minimal hours of rest.[3] In 1984, 80% of flight attendants flew 70–85 h/mo.[9]

The average age of airline pilots is 41.5 yr, and their average length of employment with their current airline is 12–13 yr, according to data from the FAA Civil Aeromedical Institute[13] and the Air Line Pilots Association (personal communication, 1985). Flight attendants have a median age of 34 yr (see Table 1-4), with 22% under 30; their average length of employment with their current airline is around 15 yr. Although 99% of pilots are men, 85% of flight attendants are women; 61% of flight attendants are married, and 43% have children.[8 9 13]

THE U.S. AIRLINE INDUSTRY

When gathering or reporting information on the nation's air carriers, FAA distinguishes between certificated route air carriers, which operate under the rules of Title 14, Part 121, of the Code of Federal Regulations (14 CFR 121)[4] and hold certificates of public convenience and necessity, and commuters or air taxis, which operate under 14 CFR 135.[1]

TABLE 1-4 Demographic Characteristics of Flight Attendants, 1985[a]

Descriptor:	Proportion of AFA Members, %
Age, yr:	
<26	7
26–29	15
30–34	29
35–39	30
40–44	13
45–49	4
50+	2
No. years employed with current airline:	
<1	10
1–5	13
5–10	22
10–15	22
15–20	23
>20	10
Race:	
Black	6
Asian	3
Hispanic	2
White	86
Other	3
Highest education achieved:	
High-school graduate	16
Some college	22
2 yr of college	24
4 yr of college	32
Some graduate school	4
Postgraduate degree	2
Average monthly hours flown, 1984:	
<70	12
70–74	18
75–79	28
80–85	34
>85	8

[a] Data from Peter D. Hart Research Associates.[9]

In 1984, there were 67 certificated air carriers; 47 were engaged in scheduled air carrier services, and the remainder provided nonscheduled (mainly charter) services. FAA classifies Part 121 airlines according to their annual operating revenues. A major airline has annual operating revenues of over $1 billion; a national airline, $75 million–$1 billion; a large regional airline, $10–$75 million; and a medium regional airline, less than $10 million.

In 1984, the 11 major U.S. airlines accounted for 67.4% of scheduled domestic enplanements and 79% of scheduled domestic revenue passenger miles (compared with 96% in 1978, before deregulation). National carriers accounted for 23.3% of scheduled domestic enplanements in 1984. The large and medium regional airlines carried 3.8% of the travelers.[11 12]

Figure 1-3 characterizes the U.S. commercial aircraft fleet by major aircraft type. Table 1-5 lists the major aircraft used by U.S. certificated air carriers projected through 1987. Wide-body aircraft (B-747, B-767, A-300, DC-10 and L-1011) accounted for 43% of the total capacity in 1984 (Table 1-6). Four aircraft models (B-727, DC-9, B-737, and DC-8) accounted for an additional 53%. The remaining seating capacity was primarily on medium or large models being phased into or out of the market and on small aircraft. Flight time and total seating capacity for U.S. airlines in 1984 are presented by aircraft type in Table 1-6.

In 1984, the approximately 175 commuter airlines carried 5.5% of the passengers and accounted for 1.1% of the total revenue passenger miles.[11 12] Of the aircraft used by commuter airlines in 1984, 77.5% had fewer than 20 seats.[11] Because small aircraft account for a very small percentage of the total revenue passenger miles flown and ventilation rates on small aircraft are generally much higher than on larger planes, this study does not address in any detail the problems associated with air quality in these aircraft.

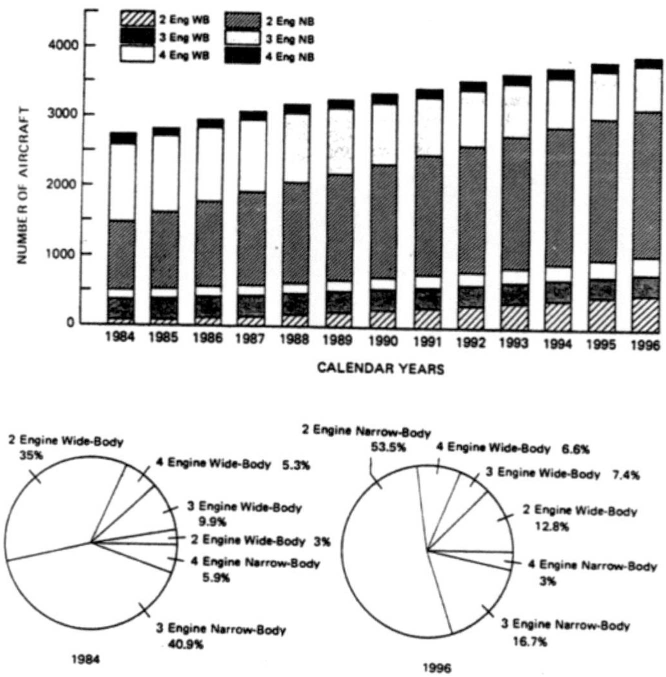

FIGURE 1-3
Large jet aircraft in U.S. commercial airline service. Reprinted from U.S. Federal Aviation Administration.[11]

TABLE 1-5 Aircraft Use, 1983–1987a

Manufacturer	Model	No. in Fleet 1983	1984[b]	1985[b]	1986[b]	1987[b]	Passenger Capacity	Total Flight Time, 1984, h	No. Engines[c]	No. Flight Crew
Airbus	A-300	34	38	38	38	38	220–375	101,143	2	3
Boeing	B-707	24	22	—	—	—	125–165	39,243	4	3
	B-727	1,122	985	985	985	985	94–145	2,990,821	3	3
	B-737	348	391				103–186	1,006,238	2	2
	B-747	146	134	134	134	134	331–550	537,142	4	3
	B-757	15	19	44	66	96	186–220	50,022	2	2
	B-767	49	53	69	72	105	211–330	172,705	2	2
British Aerospace	BAE-146	3	14	20	20	20	93–109	14,140	4	2
	BAC-111	36	33	21	21	21	74–119	59,555	2	2
Lockheed	L-1011	116	103	116	116	116	250–400	308,180	3	3
McDonnell Douglas	DC-8	133	157	50	50	50	116–259	270,728	4	3
	DC-9	557	594	616	646	665	90–135	1,438,339	2	2
	DC-10	155	174	155	160	160	250–380	487,831	3	3

[a] Data from U.S. Federal Aviation Administration.[12]

[b] Figures reflect reduction in numbers due to nonpassenger use: DC-8, B-727-200, B-747, and DC-10.

[c] All turbojet.

Flight Time and Exposure of Public on U.S. Airliners, 1984[a]

Manufacturer	Model[b]	No. in Fleet	Est. Ave. Seating Capacity	Total Flight Time, h	Seat-hours, thousands (% of total)
Boeing	B-727	1,161	120	2,990,821	358,899 (27.72)
Boeing	B-747	156	452	537,142	242,788 (18.75)
McDonnell Douglas	DC-9	594	115	1,438,339	165,409 (12.77)
McDonnell Douglas	DC-10	174	310	487,831	151,228 (11.68)
Boeing	B-737	391	120	1,006,238	120,749 (9.32)
Lockheed	L-1011	103	300	308,180	92,454 (7.14)
Boeing	B-767	53	250	172,705	43,176 (3.33)
McDonnell Douglas	DC-8	157	150	270,728	40,609 (3.14)
Airbus	A-300	38	280	101,143	28,320 (2.19)
Boeing	B-757	19	200	50,022	10,004 (0.77)
British Aerospace	BAC-111	33	100	59,555	5,955 (0.46)
British Aerospace	BAE-146	14	100	14,140	1,414 (0.11)
Boeing	B-707	22	145	39,243	5,690 (0.44)
de Havilland	DHC-7	46	50	106,287	5,314 (0.41)
de Havilland	DHC-6	107	20	176,233	3,525 (0.27)
Beechcraft	BE-99 turbo	85	15	199,205	2,988 (0.23)
Fokker	F-28	23	70	33,036	2,313 (0.18)
Fairchild	F-27	23	44	35,521	1,563 (0.12)
Fokker	F-27	14	56	25,056	1,403 (0.11)
Piper	PA-31	110	10	114,330	1,143 (0.09)
Cessna	C-402	112	6	166,914	1,001 (0.08)
Fairchild	F-227	9	48	17,053	819 (0.06)

[a] Data from U.S. Federal Aviation Administration.[12]
[b] Includes models with seat-hours greater than 0.01% of total.

FAA DATA ON SELECTED INCIDENTS

Commercial air carriers are required to report each accident or incident that involves a threat to the airworthiness of an aircraft or to the safety of passengers. The reports are recorded in the FAA Accident/Incident Data System. Table 1-7 summarizes the sources of in-flight fires and explosions, ground fires, and occurrences of cabin smoke from 1980 to 1985, and Table 1-8 summarizes emergency descents and deployments of oxygen masks in the same period (see Appendix B for complete listings from the FAA Accident/Incident Data System). Table 1-9 summarizes incidents that involved smoke or fumes in cockpits and cabins in 1974–1983, according to a separate data base, the FAA Civil Aeromedical Institute's Cabin Safety Data Bank. Emergency situations are discussed in greater detail in Chapter 4.

TABLE 1-7 In-Flight Fires and Explosions, Ground Fires, and Cabin Smoke , 1980–1985a

Source of Accident or Incident	No. In-Flight Fires or Explosions and On-Ground Fires	No. Reported Incidents of Cabin Smoke
Mechanical failure (including engines, landing gear, air-conditioning, etc.	38	30
Electric malfunction (including navigation, communication, and control instruments, etc.)	5	20
Food-service galley (except ovens and food)	2	5
Ovens or food	3	5
Passenger cabin: cigarettes and lighters	4	1
Passenger cabin: other (including lighting, projectors, speakers, etc.)	1	2
Lavatories: paper and waste chutes	4	1
Lavatories: other	5	0
Cargo compartment	1	-
Deicing malfunctions (e.g., deicing fluid)	-	3
Other	-	4
Undetermined	-	4
Total	63	75

[a] Data from FAA Accident/Incident Data System (AIDS.[14] See Appendix B.

TABLE 1-8 Emergency Descents and Deployment of Oxygen Masks, 1980–1985a

Cause of Incident	No. Incidents
Engine malfunction (e.g., bleed airports, turbine failure, power loss, fire warning system, and fire)	8
Ventilation system failure (e.g., air-conditioning turbine, ducting, outflow valve, and water separator)	14
Control equipment malfunction (e.g., electric panel, fire prevention system, pressure controller, and broken wire)	23
Landing gear malfunction (e.g., hydraulic fluid loss)	1
Medical (e.g., passenger illness or flight crew member unconscious)	2
Other (e.g., hydraulic flap failure, burst water tank, wiring pylon cracked, bird in valve, fuselage skin fatigue, and bomb threat)	6
Unknown or unclassified	17
Total	71

[a] Data from the FAA Accident/Incident Data System (AIDS).[14] See Appendix B.

TABLE 1-9 Incidents Involving Smoke or Fumes in Cabin or Cockpit, 1974–1983a

Year	No. Incidents with Emergency Landing	No. Incidents with No Emergency Declared	No. Incidents with Unknown Landing Status	Total No. Incidents
1974	9	2	7	18
1975	11	1	4	16
1976	14	1	7	22
1977	11	8	2	21
1978	13	2	2	17
1979	19	3	2	24
1980	17	4	0	21
1981	9	7	1	17
1982	10	5	5	20
1983	17	9	4	30
Total	130	42	34	206

[a] Data from Higgins.[7]

REFERENCES

1. Air taxi operators and commercial operators. Code of Federal Regulations, Title 14, Pt. 135. Washington, D.C.: U.S. Government Printing Office, 1985.
2. Air Transport Association of America. Air Transport 1985: The Annual Report of the U.S. Scheduled Airline Industry. Washington, D.C.: Air Transport Association of America, 1985.
3. Association of Flight Attendants. Petition of Proposed Rule Making. 1985. (unpublished)
4. Certification and operations: Domestic, flag and supplemental air carriers and commercial operators of large aircraft. Code of Federal Regulations, Title 14, Pt. 121. Washington, D.C.: U.S. Government Printing Office, 1985.
5. Flight time limitations: All flight crewmembers. Code of Federal Regulations, Title 14, Pt. 121.471. Washington, D.C.: U.S. Government Printing Office, 1985.

6. Gallup Organization, Inc. 1985 Air Travel Survey. Washington, D.C.: Air Transport Association of America, 1985.
7. Higgins, E. A. Protective breathing: Oxygen mask use/problems, pp. 61–65. In Flight Safety Foundation, Inc. Proceedings of Cabin Safety Conference and Workshop, December 11–14, 1984. Washington, D.C.: U.S. Federal Aviation Administration, Office of Aviation Safety, 1985.
8. Joint Council of Flight Attendants Unions. About the Joint Council. (unpublished communication, 1985)
9. Peter D. Hart Research Associates, Inc. Membership Survey, Association of Flight Attendants. 1985. (unpublished)
10. U.S. Bureau of the Census. Statistical Abstract of the United States, 1985. 105th ed. Washington, D.C.: U.S. Government Printing Office, 1984.
11. U.S. Federal Aviation Administration. FAA Aviation Forecasts, Fiscal Years 1985–1996. FAA-APO-85-2. Washington, D.C.: U.S. Federal Aviation Administration, 1985.
12. U.S. Federal Aviation Administration. FAA Statistical Handbook of Aviation, Calendar Year 1984. Washington, D.C.: U.S. Federal Aviation Administration, 1985.
13. U.S. Federal Aviation Administration, Civil Aeromedical Institute, Aeromedical Certification Branch. 1984 Aeromedical Certification Statistical Handbook. AC 8500-1. Washington, D.C.: U.S. Federal Aviation Administration, 1985.
14. U.S. Federal Aviation Administration, Safety Analysis Division. Analyses Performed 3 Dec. 1985 and 6 Jan. 1986 Using Accident/Incident Data System (AIDS) for Committee on Airliner Cabin Air Quality. (computer printouts)

15. U.S. Travel Data Center. National Travel Survey: 1984 Full Year Report. **Washington, D.C.**: U.S. Travel Data Center, 1985.

2

Environmental Control Systems on Commercial Passenger Aircraft

Although the variety of airplanes operating throughout the world is large, the basic designs of the environmental control systems (ECSs) used on most aircraft in commercial service are remarkably similar. In simplified terms, air is first compressed to high pressure and temperature and then conditioned in an environmental control unit (ECU), where excess moisture is removed and the temperature necessary for heating or cooling the airplane is established. The conditioned air is then delivered to the cabin and cockpit to maintain a comfortable environment.

DESCRIPTION OF ENVIRONMENTAL CONTROL SYSTEMS

Compressed-Air Sources

On the ground, compressed air for the ECS can be obtained from an auxiliary power unit (APU), a special ground cart (GCU), airport high-pressure hydrants, or the aircraft engines. In flight, however, compressed air is obtained almost exclusively from the compressor stages of the aircraft engines.

In most respects, the composition of ambient outside air will not be changed in the compression cycle. Contaminants will in general be neither removed nor added. Some particles can be removed by centrifuging in the port through which air is extracted from the engine. One contaminant that can be affected by the heat of compression is ozone. In supersonic flight of the Concorde, the compressed-air temperatures are so high that nearly all the ozone is destroyed in the engine, and no further treatment with catalysts or filters is needed. In all other commercial aircraft, the normal temperature of the compressed air taken from the engine for air-conditioning is not adequate to reduce the free ozone concentration substantially.

Oil seal leaks have sometimes permitted engine oil to leak into the compressors, and oil can then enter the bleed air in the form of vapor or, in extreme cases, mist. In recent years, oil seal failures have not been a problem. Where engine seal design does not prevent oil vapors from entering the system, turbo-driven or engine-driven compressors are installed. The use of separate compressors increases weight, decreases reliability, and imposes additional maintenance requirements.

For ground air-conditioning, high-temperature compressed air can be supplied to the cabin through the ECU from an onboard APU or from a portable ground cart. These units operate much like the main engines in generating compressed air; however, the design is usually optimized for efficient delivery of compressed air, rather than propulsive thrust. The air supplied by these units is taken from the ramp area and contains whatever contaminants are present in that area.

High-pressure air can also be supplied from airport facilities. Because of the lower operating cost of fixed electrically driven generating and compressor units and the reduction in ramp contamination and noise, the use of high-pressure ground air facilities is increasing.

Preconditioned low-pressure air, which is the lowest-cost source of heating and cooling, can be supplied directly to the airplane air distribution system through ground connections from portable air-conditioning units or from central airport facilities. The air supplied is taken from the ramp area or the terminal and contains contaminants typical of those areas.

The Environmental Control Unit

In flight, high-pressure, high-temperature air is conditioned by processing through the ECU before delivery to the cabin. The ECU (or "pack") usually consists of an air-cycle machine (ACM) and one or more heat exchangers.

A simplified schematic, Figure 2-1, shows how air is conditioned in cruise and delivered to the cabin to meet heating, cooling, ventilation, and pressurization requirements.

Normally, ambient temperatures at cruise altitudes are low enough for bleed air to be cooled adequately by the heat exchangers alone, and the ACM is completely bypassed. On the ground, at lower altitudes, in "hot-day" conditions, and during low-speed flight, the ACM will be used to cool the air further to meet cabin requirements.

Mechanical water separators are used for ground and low-altitude operation to remove water droplets from the outside air. These droplets are formed when the air is expanded and cooled in the ECU turbine and are very fine, about 5 μm in diameter. The mechanical water separator contains a bag (usually of finely woven Dacron or frayed Teflon) that coalesces the fine droplets and permits them to be centrifuged out in the downstream section of the water separator.

The efficiency of the water separator generally is 80–90%. Water not removed enters the cabin ducting, where it absorbs heat from the distribution system and is vaporized. The liquid droplets sometimes appear as fog emanating from the outlet grilles.

To prevent freezing of the water separator, ACM discharge temperatures must be limited to about 35°F.

Recent developments have led to the use of high-pressure water separators that condense and remove moisture from the bleed air before it expands in the turbine. This design, which is currently in use on B-757s and B-767s,[18] permits the moisture content of air entering the turbine to be less than 15 grains/lb (2 g/kg), which in turn permits the ECU to discharge air from the turbine at temperatures well below freezing. If air were introduced into the cabin at subfreezing temperatures, draft and local cold areas would be created. Therefore, recirculated cabin air is mixed with the cold ECU discharge air at a ratio of about 1:1 to achieve a minimal temperature of 35–40°F, the minimal temperature to prevent icing. Through this mixing and operation of the ECUs to produce very low discharge

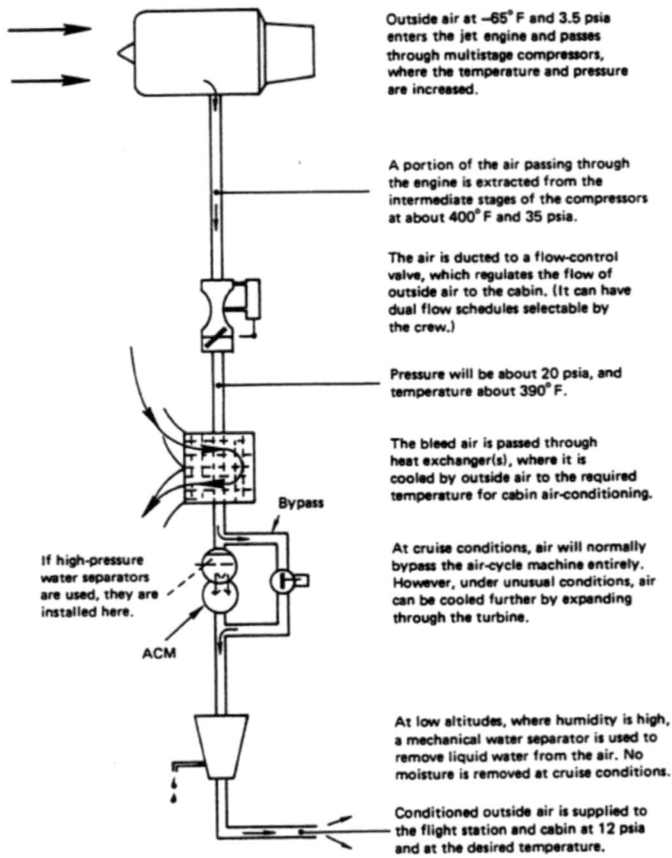

FIGURE 2-1
Operation of aircraft environmental control unit in cruise conditions at 35,000 ft.

temperatures, the cooling capacity of each pound of outside air is almost doubled, compared with that in systems that use conventional water separators.

Air Distribution

Outside air, conditioned by the ECU to the proper temperatures, is usually mixed in a plenum and then distributed to the cockpit and the cabin zones. A large, wide-body aircraft might have as many as six individual temperature-controlled zones, each with its own supply ducting system, whereas a smaller, narrow-body aircraft usually has only two such zones, one for the cabin and one for the cockpit. Airflow to each zone is established by the cooling requirement of the zone. The total cooling requirement is met by supplying a quantity of air to the zone at the low-temperature limit (40°F). Because passenger and crew heat loads account for only 40–50% of the total cooling requirement—whereas the remaining 50–60% (lighting, solar loads, and conduction through cabin structure) is determined by cabin areas, rather than by number of passengers—outside air will not be distributed strictly on a per-passenger basis. First-class and business sections of the cabin might have 2–3 times as high a ventilation rate per occupant as the economy section.

Because of the larger solar and electronic cooling loads in the cockpit, ventilation per flight crew member might be 10 times as high as that in the cabin, or even higher.

The distribution of outside air (or outside and recirculated air) to the cabin is usually fixed by the ducting design and flow-balancing orifices. However, some combi-aircraft (aircraft modified to carry passengers and cargo in the main cabin) have provisions to reduce airflow to the aft section when cargo is carried in that section of the cabin. Individual air outlets (or gaspers) that can be adjusted by the passenger for air flow and direction can be supplied with cold air taken from the ECU discharge or with air from the main supply ducts in the cabin. Thus, the air can be fresh, a mixture of fresh and recirculated air, or, as in the case of the DC-10 seat-mounted gaspers, only recirculated air. Gaspers are generally being phased out in the newer and wide-body aircraft.

The main supply air enters the cabin through fixed outlets, which can be in the ceiling or in the sidewalls below the overhead storage bins. Some aircraft have both types of outlets, and the selection of a system to use is based on whether the aircraft is being heated or cooled.

Exhaust Systems

Air is normally exhausted from the cabin through floor-level grilles, which run the length of the cabin on both sides along the sidewall. The exhaust air is directed alongside or through the lower-lobe cargo compartments, where it can provide some heating or cooling. The air is then exhausted overboard through outflow valves controlled to maintain the desired cabin pressure. Figure 2-2 illustrates typical passenger cabin airflow patterns. Cabin exhaust air is also used to cool avionics and electric equipment and then discharged overboard through the outflow valves.

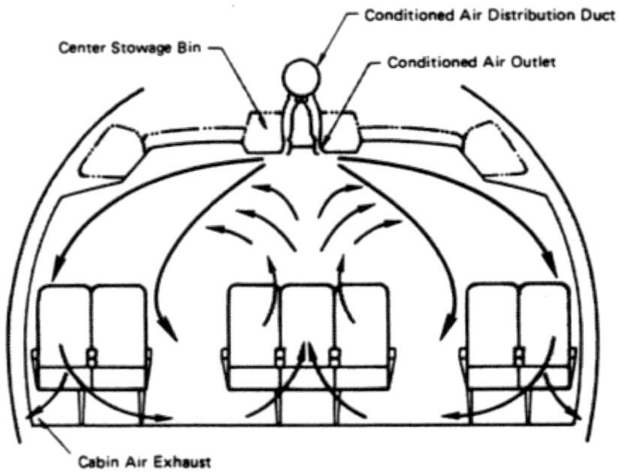

FIGURE 2-2
Typical passenger cabin airflow patterns. Reprinted from Lorengo and Porter.[9]

Lavatories are ventilated with cabin air drawn through them. About 3–5 cfm is supplied with an individual air outlet. The ventilation air is exhausted overboard, either directly through a port in the skin of the aircraft or through ducts that direct the air toward the outflow valves.

Air is exhausted from galleys to exhaust moisture and food odors and to prevent their diffusion into the cabin. Galley ventilation air can be ducted directly overboard or to the outflow valves. Galleys and lavatories are often exhausted through a common duct system.

Recirculation Systems

Recirculation systems have been used on the early Convair 880 and 990, B-707, DC-8, Lockheed Electra, and many other older aircraft that used vapor-cycle cooling systems. The use of air recirculation systems in modern aircraft has recently increased with the advent of higher engine bypass ratios, higher jet-fuel costs, the design of "stretched" versions of production aircraft, and the development of advanced ECUs that use high-pressure water separators. The bypass ratio is the ratio of fan air flow to high-pressure or engine-core air flow. The fuel and performance penalties associated with bleed air extraction increase as the bypass ratio increases. As aircraft are "stretched" to increase seating capacity, recirculation systems are added to improve air distribution and circulation. To use the greater cooling capacity of ECUs equipped with high-pressure water separators, warm cabin air must be mixed with outside air to raise the temperature of air supplied to the cabin. The very cold ECU discharge air would cause condensation and local draft problems if introduced into the cabin without mixing.

In 1985, about 30% of the seat-hours flown by U.S. airlines were on aircraft with recirculation systems. By 1990, this percentage will have increased to 40%, as more of the newer, fuel-efficient aircraft enter service.

Air for recirculation can be taken from the general space above the ceiling (B-747), from slotted openings in the ceiling (DC-10), or from underfloor spaces. In about 75% of the aircraft with recirculation systems in 1985, the recirculated air was returned to the same zone. In the remaining 25%, recirculated air was mixed with outside air and distributed throughout the cabin and, in the B-767 and some models of the B-757, to the cockpit. By 1990, the percentage in which air is totally mixed with outside air will increase from 25% to about 45%.

Recirculation air can be filtered to remove lint, aerosols, and gaseous tars from tobacco smoke. Although the technology is well developed for removing gases with charcoal, only some models of the B-757 and B-767 currently use this method in the recirculation system. These aircraft have charcoal filters available as an airline option.[18] Particle filters that remove particles as small as 0.3 μm are installed in 80% of the aircraft with recirculation systems.

Some aircraft manufacturers and filter manufacturers are conducting research to improve equipment for removing particles and gases from recirculated air. Programs begun in 1985 are investigating the use of electrostatic precipitators in aircraft to remove particles (McDonnell Douglas, personal communication, 1985; Boeing, personal communication, 1985).

Temperature Control

Temperature in each zone of the aircraft is controlled to a value selected by the flight crew, usually between 65 and 85°F. Turbine bypass and heat-exchanger airflow valves are typically used to establish the ECU discharge temperature and a zone reheating system to establish supply temperatures for each zone. Where discharge air from the ECUs is mixed in a plenum, the ECU discharge temperature is controlled to meet the demands of the zone that requires the coldest air, and a reheating system is used to add hot bleed air to the other zones, which need less cooling or more heating. Operation of the zone reheating system does not substantially affect air flow and distribution to the zones.

Pressure Control

An automatic pressure control system establishes the cabin pressure as a function of altitude and controls the rate of change of cabin pressure during climb and descent. Cabin pressures during cruise are based on the allowable pressure difference between the cabin and the outside. The allowable difference varies with aircraft design and is a structural limit. Figure 2-3 shows the relationship between cabin and flight altitudes for typical commercial aircraft. The maximal cabin altitude cannot exceed 8,000 ft for normal operation up to the certified aircraft altitude.

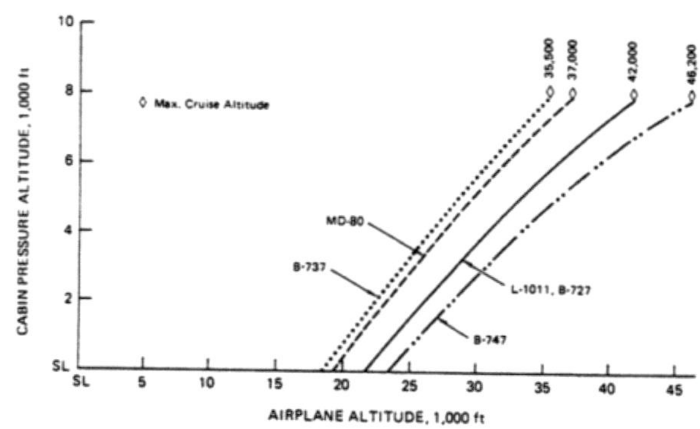

FIGURE 2-3
Cabin-pressure altitude at maximal differential pressure.
SL = sea level. Data from Lorengo and Porter.[9]

The rate of pressure change is controlled during climb and descent to meet criteria for passenger comfort and pressure-difference limits of the aircraft. The recommended rates of change of pressure for passenger comfort are 500 ft/min (-0.256 psi/min) during climb and 300 ft/min (+0.154 psi/min) during descent.[12]

The crew can select higher or lower rates of change, but the controls are normally set at the recommended value, which is usually identified by an index mark on the pressure control panel.

PERFORMANCE OF ENVIRONMENTAL CONTROL SYSTEMS

A major aspect of aircraft ECS performance, and one that was considered of primary importance by the Committee in the study of cabin air quality, is the ventilation rate. Aircraft ventilation rate is defined as the amount of outside air supplied to the passengers and crew in cubic feet per minute per occupant and is determined by dividing the outside air supplied at design conditions by the passenger and crew seats. Normal conditions include full passenger load, operation of all ECUs at rated flow, and steady-state cruise. Airlines may increase the passenger capacity above what is shown in Table 2-1, and that would reduce passenger ventilation rates. In addition, the operation of the ECU can affect the ventilation rate. Minimal airflow must be adequate to meet heating, cooling, and

TABLE 2-1 Effect of Flow Options and Seating Density on Passenger Ventilation Ratea

		ECU Operation		Outside Air Per Passenger, cfm		
Aircraft Model	No.Passengers	No.	Flow Schedule	First Class	Overall Average[b]	Economy Class
DC-10	290	3	High	—	18.5	—
	290	2	Low	—	6.9	—
	274	3	High	—	21.0	—
	274	2	Low	—	8.4	—
L-1011-1	279	3	Normal	128.7	—	19.5
	279	2	Normal	17.3	—	11.6
L-1011-500	235	3	Normal	51.0	—	18.6
	235	2	Normal	30.5	—	11.2
B-747-200	381	3	Normal	40.4	—	17.1
	381	2	Normal	26.5	—	8.5
	300[c]	3	50%	11.4	—	10.6
	265[c]	2	50%	12.7	—	10.6
B-767-200	217	2	Normal	10.0	—	9.9
	217	2	Optional filters installed	20.0[d]	—	19.8[d]

[a] Based on data from Aerospace Industries Association of America.[1]

[b] Section data not available.

[c] Recommended maximal number of passengers (Boeing).

[d] Includes treated air.

pressurization requirements throughout the aircraft. Variation in seating density between first-class and economy sections causes a variation in the outside-air ventilation rates in these areas.

If the ECS is made up of three independent ECUs, the operator might be permitted to dispatch the aircraft with one ECU inoperative or, if the aircraft was dispatched with all ECUs operating, to shut down one of the units. In either case, the ventilation rate can be less than originally specified by the manufacturer.

Crew options also include selection of individual ECU flow rates. High and low flow schedules are sometimes incorporated into the ECU flow control valve, to permit crew operation of each unit at normal or reduced flow. Reduced-flow schedules are usually one-half to two-thirds of normal. Operators of the B-747 and DC-10 also have access to dual-schedule flow control valves that permit selection of ventilation flow in increments of less than one full ECU. This design is available as an airline option. The option of reducing flow by shutting off a pack is now available only on the B-747, DC-10, and L-1011 aircraft, all of which have three independent ECUs.

Because in normal cruise conditions the ECUs have more than adequate heating and cooling capacity, ventilation can be reduced with no substantial effect on cabin temperature or pressure. Airlines are therefore financially motivated to save fuel by reducing the amount of ventilating air that is taken from the engines. A NASA-sponsored study in 1980[10] showed that about 62,000 gal of fuel, or about 1% of the annual total, could be saved per year per DC-10 if the flight crew reduced the ventilation flow from 18 cfm to 8 cfm per passenger.

The combined effect on passenger ventilation rate of reducing ventilating air flow and variations in seating density is shown in Table 2-1.

Load Factors

The load factor, or percentage of aircraft seats occupied, is a vital statistic in airline operation. A measure of potential profitability, it is used by airlines in route analysis, equipment assignment, and decisions regarding purchase of new equipment.

Load factor also has a major effect on the ventilation rate of a flight. Just as the number of seats cannot be tailored to each flight, the ventilation system on newer aircraft has only limited variability. Thus, the ventilation rate is much higher on low-load flights than when the aircraft is full.

The average load factor in the United States declined in fiscal 1984 to 57.8%. FAA projects a rise to 59.1% in 1985 and, after a slight decline in 1986, a slow rise to 63.8% in 1996.[19]

The effects of load factor on ventilation rate and resulting air quality are to a large extent buried in the averaging process. To measure and evaluate ventilation rates, it is necessary to examine individual flights. An ATA study of individual flight data that had been collected by CAB in 1975–1976 showed that the frequency distribution of load factors can be well represented by a normal curve at the lower end. The upper-end or right hand extension of the curve is cut off by the physical limit on the available seats. The extension of the normal curve past the 100% load factor is called "unaccommodated demand" by ATA.[3] Unaccommodated demand occurs when the number of requests for passenger seats is greater than the capacity. Therefore, although passenger demand follows a normal distribution, the flight load-factor distribution is a truncated normal curve, as shown in Figure 2-4.

The Boeing Commercial Airplane Company studied the relationship between average passenger load factor and unaccommodated demand (the percentage of passengers who cannot be accommodated at their desired departure times) and developed a program that defines this relationship.[3] The relationship is shown in Table 2-2. On the basis of average load factors,[19] the resulting unaccommodated demand (Table 2-2), the truncated normal distribution curve (Figure 2-4), and the assumption that the unaccommodated demand will represent the percentage of flights at 100% load factor, a load-factor distribution histogram was prepared for the years 1985 and 1990. It is shown in Figure 2–5. A ventilation distribution was then calculated by the Committee on the basis of the load factors in Figure 2–5, the ventilation rate of each major aircraft model, and the percentage of seat-hours flown by that model (Table 2–3). These ventilation distributions (Figures 2–6 and 2–7) were calculated as follows: A ventilation rate multiplying factor (MF), based on a load factor, was used to modify the basic ventilation rate for each aircraft type (Table 2–3). For a load factor of 50%, MF = 2; for a load factor of

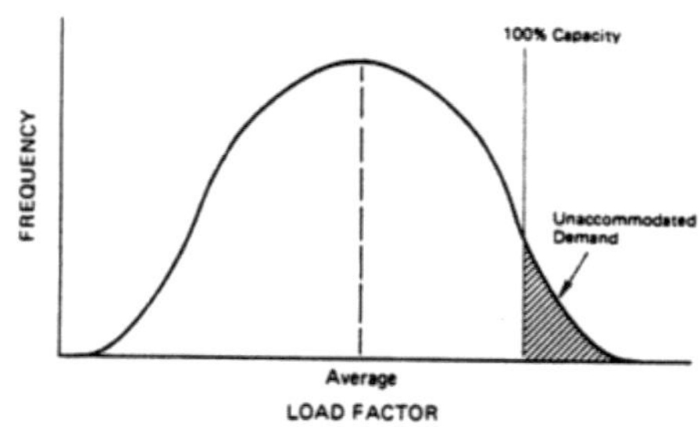

FIGURE 2-4
Demand distribution at various load factors. Adapted from ATA.[3]

TABLE 2-2 Unaccommodated Demand vs. Load Factora

Average Load Factor, %[b]	Unaccommodated Demand, %[c]
80	64
70	21
60	6
50	2

[a] Based on data from ATA.[3]

[b] Average year-round load factor for U.S. airline industry.

[c] Percentage of passengers who cannot be accommodated at their desired departure times.

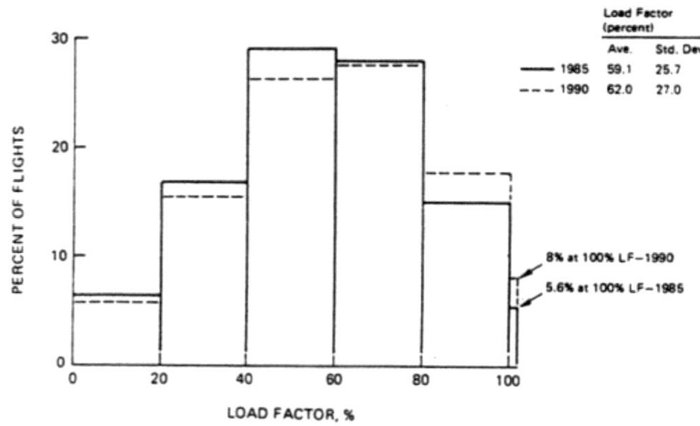

FIGURE 2–5
Load-factor distribution. Based on data from U.S. FAA19 and ATA.[3]

TABLE 2–3 Seat-Hours Flown and Ventilation Rates, by Aircraft Typea

	Proportion of Total Seat-Hours[b] Flown Annually, %				
Aircraft Model	1984	1985[c]	1986[c]	1987–1990[c]	Outside Air, cfm/occupant[d]
B-727	27.7	24.8	24.2	22.9	17.5
B-747	18.8	19.5	19.1	18.0	8.5[e]-40.4
DC-9/MD-80	12.8	16.0	16.2	15.9	13.7
DC-10	11.7	10.8	10.9	10.4	6.9[e]-18.5
B-737–200	8.4	8.0	7.9	7.7	15.3
L-1011	7.1	9.1	8.9	8.5	11.6[e]
B-767–200	3.3	3.4	3.3	4.0	9.9[e]-18.8[f]
A-300	2.2	2.5	2.4	4.0	14.3
DC-8	3.1	2.2	2.0	1.9	13.0
B-757	0.8	1.8	2.7	3.8	8.8[e]-18.7[f]
B-737–300	0.4	1.8	2.4	2.9	8.6

[a] Based on data from Lorengo and Porter.[9]

[b] Seat-hours=(number of seats installed)(flight duration, hours).

[c] Projection.

[d] At 100% passenger load factor, including normal complement of cabin crew.

[e] At minimal flow.

[f] Incorporates optional particle filters and charcoal filters to remove gaseous and aerosol contaminants at 90% efficiency. Higher flow rate shown is with charcoal filters installed.

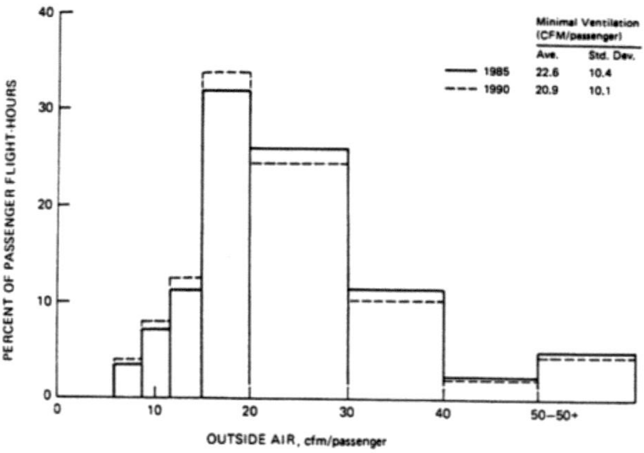

FIGURE 2-6
Ventilation rate distribution, minimal flow, for major U.S. domestic airlines. Passenger flight-hours = (number of passengers) (flight duration, hours). Based on data from U.S. FAA19 and ATA.[3]

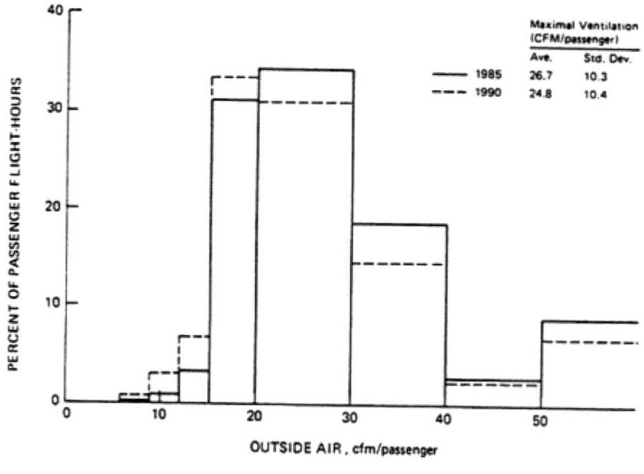

FIGURE 2-7
Ventilation rate distribution, maximal flow, for major U.S. domestic airlines. Passenger flight-hours = (number of passengers) (flight duration, hours). Based on data from U.S. FAA19 and ATA.[3]

70%, MF = 1.43. The load-factor frequency was taken from Figure 2–5, with the percent unaccommodated demand arbitrarily assigned to 100% load factor. The actual ventilation rate (AVR) then was summed for each aircraft type, on the basis of the percent of seat hours flown by that aircraft and the load-factor frequency. For example, in 1984, B-727s flying 27.7% of the total U.S. fleet seat-hours would dispatch 16.6% of the flights with a load factor between 20 and 40%. The load-factor mean of 30% was used, the multiplier was 3.3, and the AVR was 57.8 cfm/passenger. The total number of seat-hours at 57.8 cfm/passenger then was 0.166 x 0.277 = 0.046 (4.6%). To convert seat-hours to passenger-hours, this value was multiplied by the load factor for this segment (30%). Thus, B-727s provided 1.38% of passenger flight hours, with an AVR greater than 50 cfm/passenger. The values for each airplane and each load factor segment (Figure 2-5) were summed to generate Figure 2-6. Figure 2-7 was generated in the same way, except that minimal ventilation rates were used. The ventilation rates used in preparing Figure 2-6 were based on the flight crew's use of minimal flow permitted by the aircraft design. The frequency of use of low-flow options by flight crews is unknown. The effect of crew use of maximal flow on ventilation rate is shown in Figure 2-7. However, the trend toward lower ventilation rates is expected to continue. This will occur through the addition of recirculation systems to the existing fleet, the increased use of low-flow options, and the introduction into the U.S. airline fleet of more aircraft that use higher percentages of recirculated air (B-767, B-757, B-737-300, and MD-80).

EFFECT OF VENTILATION ON TOTAL CABIN ENVIRONMENT

Outside-air ventilation is the prime variable affecting contamination in the aircraft cabin. At high outside-air ventilation rates, passenger well-being is increased with respect to carbon monoxide and carbon dioxide, contamination due to smoking, and odor. Increasing total cabin airflow (with either outside or recirculated air) also increases movement of air, which creates a feeling of freshness and reduces temperature stratification.

Higher outside-air ventilation rates lower cabin relative humidity. In addition, when the aircraft is operating in regions of high ambient ozone, cabin ozone is also increased by the increased use of outside air. An increase in total cabin airflow, with either outside or recirculated air, creates a potential for local high velocities and drafts, adds a direct fuel cost, and potentially involves costs of equipment weight and maintenance.

Ventilation and Contamination

Cabin ventilation provides air for dilution of contaminants and supplies oxygen for passengers and crew. As shown in Table 2-1 and Figures 2-6 and 2-7, outside-air ventilation rates can vary widely. Oxygen requirements for sedentary adults can be met with only 0.24 cfm.[4] Thus, even at the lowest ventilation rates on aircraft, there is no significant reduction in the percentage of oxygen in the cabin. Contamination with carbon dioxide varies inversely with ventilation rate, because carbon dioxide production by passengers is nearly constant. However, the amount of contamination with tobacco smoke (carbon monoxide and particles) depends on ventilation rate, number of smokers, and smoking rate.

Smokers on airplanes are estimated to make up 33% of the total passenger load. The average smoking rate has been estimated at 1.25-2.2 cigarettes/h per smoker. Halfpenny and Starrett[7] measured 1.25 cigarettes/h per smoker on 33 2-h flights. Cain et al.[5] used a rate of 2 cigarettes/h per smoker in 1982 odor studies, and Thayer[16] calculated an average smoking rate of 2.2 cigarettes/h on the basis of the total number of cigarettes produced, 33% of the population aged 18 and over being smokers, and a 15-h smoking day.

With a generally constant smoking rate, the concentration of tobacco smoke depends on the flow of outside air into the cabin. Passengers perceive tobacco-smoke contaminants in the form of odor and irritation of eyes and nasal passages. Acceptance of air contaminated with tobacco smoke has been measured in juries of smokers and nonsmokers in odor test rooms and in an airplane mockup. The results of three studies are shown in Figure 2-8. The difference in jury acceptance of contamination shown in Figure 2-8 is due to the evaluation criteria used by the investigators. The results obtained by Cain et al.[5] were based on odor evaluations by active smokers, and the high degree of acceptance by the occupants, compared with that reported by the nonsmoking visitors, represents odor adaptation. Halfpenny and Starrett[7] and Thayer[16] evaluated odor and occupant irritation. Because people do not adapt to the irritants in tobacco smoke—rather, the degree of irritation increases with duration of exposure, reaching a peak after about 15 min and then remaining relatively constant[7]—the acceptance of odor and irritation shown is lower than acceptance of odor alone.

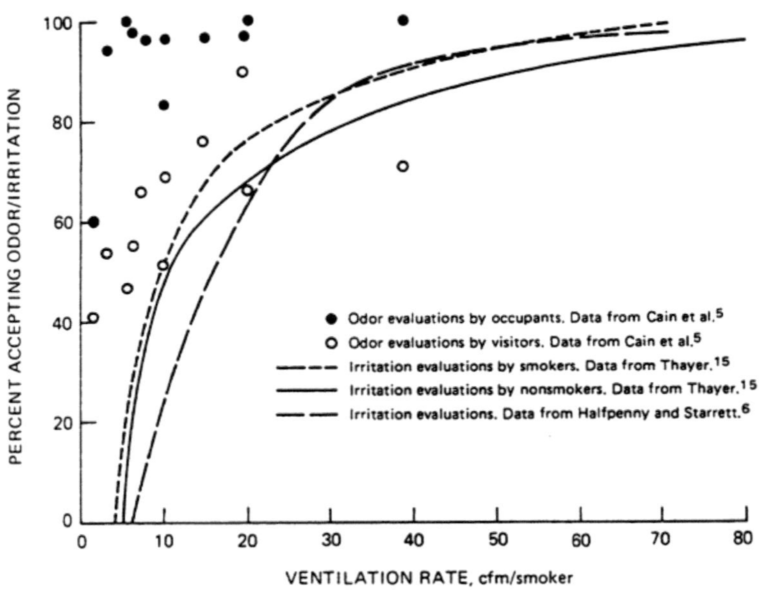

FIGURE 2-8
Relationship of ventilation rate to acceptability by smokers and nonsmokers of tobacco smoke odor/irritation. The Cain et al. data—outside-air flow (L/s) and number of cigarettes smoked—are converted to cfm/smoker, according to [(L/s)(2.118)]/[(cigarettes/h)(2)]. In their studies, the air had 50% relative humidity. Data from Cain et al.,[5] Halfpenny and Starrett,[7] and Thayer.[16]

All the data shown in Figure 2-8 were taken at relative humidities of 30–75%, which are much higher than are normally encountered in airplanes. Kerka and Humphreys[8] showed that, in general, increased humidity tended to decrease sensory response to odors and irritants. Cain et al.[5] showed that "high humidity" (75%) generated a more intense odor response than "moderate humidity" (50%). However, the degree to which low humidity typical of aircraft cabins (usually 5–10%) can affect response to odor and irritation has not been investigated.

The contamination at various ventilation rates encountered in airplane smoking sections and the average contamination in the cabin when air in smoking and nonsmoking sections is fully mixed are also shown in Figure 2-8.

Contamination in the form of tars can affect aircraft systems where cabin air is used for cooling. Avionics components that are usually cooled by cabin air are adversely affected by a buildup of tars and lint, which reduces component cooling. Particularly vulnerable are temperature control sensors that respond to a flow of cabin air. Tars and lint cause slow sensor response, which results in unstable cabin temperatures. Axial-flow fans have become so contaminated with tobacco tars that fan blades are stuck to the housing, causing motor overheating and premature bearing failures. The actual increase in maintenance costs due to tobacco smoke was not available; however, it is generally felt by airliner maintenance personnel that they are significant.[15]

Air Velocity and Cabin Flow Patterns

Circulation of air in the passenger area at velocities of 10–60 ft/min is necessary to prevent local stagnation and temperature stratification. A minimal velocity of 10 ft/min (0.05 m/s) is necessary to avoid the sensation of stagnation, whereas velocities above 60 ft/min (0.3 m/s) can create a draft sensation on the neck.[13] Aircraft distribution systems normally provide adequate circulation when the ECS is operated at full rated flow. However, when total outside air is reduced and there is no compensating recirculated air, stagnation can be created, and normal flow patterns in the cabin can be affected. Operating with reduced outside-air flow sometimes causes air from the smoking areas to be drawn into nonsmoking areas. This can occur if bleed flow is reduced to the point where controlled exhaust through outflow valves is very low and the bulk of the exhaust is through leakage paths. This can create fore and aft flow in the cabin which can spread tobacco smoke into nonsmoking zones.

Relative Humidity

Relative humidity in aircraft cabins in cruise is seldom controlled and depends entirely on the moisture given off by passengers and crew in the form of respiratory vapor and perspiration. The amount of moisture given off depends on the extent of activity and cabin temperature. A sedentary passenger normally emits about 0.7 g/min, and a cabin crew member, about 2 g/min. Because outside air is essentially dry (moisture at less than 100 ppm), cabin relative humidity varies inversely with ventilation rate (see Figure 2-9).[12]

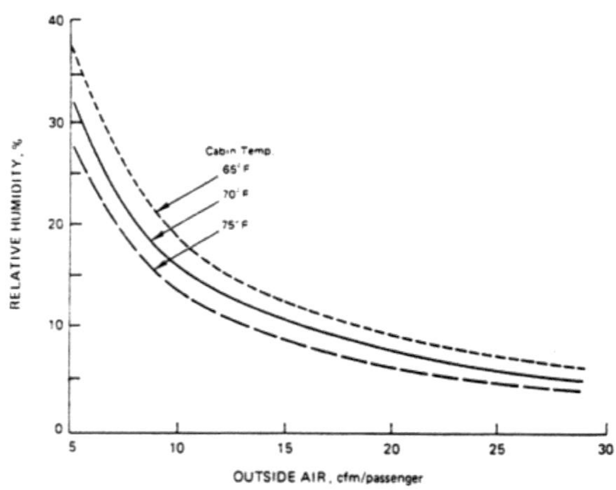

FIGURE 2-9
Relation of relative humidity and outside-air ventilation rate. Equivalent cabin altitude, 6,500 ft. Data from SAE.[12]

Ozone

Ambient ozone is present above the tropopause, whose height varies with latitude and season. It normally exists at an approximate altitude of 11 km (36,000 ft) in the middle latitudes in summer. Ozone enters the cabin with outside air through the engines and ECU. Residual cabin ozone concentration is a function of the outside concentration, the design of the air distribution system, the use of catalysts or adsorbers, and the total airflow. Each airplane has a characteristic cabin ozone retention factor, which is the ratio of the ozone concentration in the cabin to the ozone concentration in outside air after it has passed through the ECU. Normally, the retention ratio is from 0.75:1 to 1.00:1 without any recirculation, but it can be as low as 0.4:1 with recirculation.[20] Where the retention ratio is too high to limit cabin ozone to the FAR 121 maximum, alternative treatment of the outside air is required. Noble-metal catalysts are used to remove a portion of the ozone before it enters the cabin. These units have removal efficiency of 90–95%.[6] (See Chapter 5 for additional details on ozone.)

Effect of Recirculation on Contamination

Cabin recirculation systems on most airplanes result in partial or complete mixing of air in the smoking and nonsmoking sections. Recirculation air is often taken from a plenum near the outflow valve where exhaust air from all cabin sections is collected and then distributed to all sections and in some cases to the cockpit. This negates to some extent the nonsmoking/smoking sectioning of the cabin. The flow model developed by the Committee has been used to evaluate contamination in all sections as a result of recirculation designs (see Appendix A).

Cost of Ventilation

The direct cost of supplying outside air to passengers and crew includes the loss of aircraft thrust due to the extraction of high-pressure air from the engine compressors, the power loss due to the extraction of fan air for precooling, and the ram drag incurred in ECU heat-exchanger cooling. All this power loss must be compensated for by increasing engine power settings, which increases fuel consumption.

The net cost of ventilation is reduced somewhat by the use of thrust-recovery exhaust valves, which discharge exhaust air aft and produce positive thrust.

The weight penalty for basic ECS equipment should not be charged to the design ventilation air flow, because the equipment is normally sized to meet design cooling requirements, which are based on hot-day conditions at sea level. However, if the ventilation rate were increased above the flow required for cooling as designed, then the weight penalty of the added ECS equipment (large ducts, valves, heat exchangers, etc.) would constitute an added ventilation cost.

Studies by aircraft manufacturers to establish ventilation costs have shown significant variation in those costs. The Boeing Commercial Airplane Company estimated a fuel-burn penalty of 0.015 gal/h per cubic foot per minute (gph/cfm) for the B-727 and B-747,[11] whereas McDonnell Douglas estimated 0.009 gph/cfm for a DC-10 in a NASA-funded fuel-reduction program.[10] These variations are due in part to the stage length used in the analyses and the ambient conditions; fuel penalty is higher in climb and on hot days. The greatest variation, however, is due to the drag coefficients used.

The range of fuel costs in gph/cfm per passenger based on these analyses is shown in Figure 2-10. To place the cost of aircraft ventilation in perspective, it can be compared with the cost of providing equivalent fresh air in commercial or residential buildings. The cost of providing outside air for an airplane is 22–37 times the cost of providing the same amount of air in Washington, D.C., during the coldest month, January.[17]

Fuel costs constitute a substantial percentage of operating costs. At the current price of 76–86 cents/gallon, fuel costs for the wide-body fleet (B-767, B-747, A-300-B4, DC-10-10, and L-1011) in the quarter ended September 30, 1985, ranged from 52 to 68% of the cash operating cost and from 37 to 57% of the total aircraft operating expenses.[2]

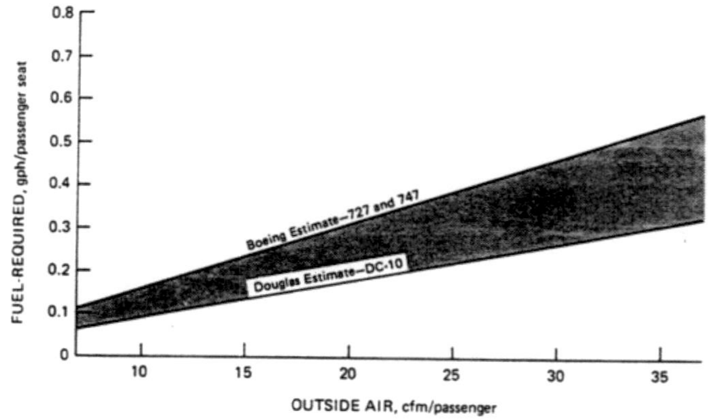

FIGURE 2-10
Fuel required for ventilation with outside air. Data from Reese.[11]

REFERENCES

1. Aerospace Industries Association of America, Inc. Airplane Air-ConditioningSystem Configuration and Air Flow Data for Selected Boeing, Douglas, and Lockhead Aircraft. (unpublished document, September 17, 1985)
2. Aircraft operating data. Air Transport World 23(5):188, 1986.
3. Air Transport Association of America, Economics and Finance Department. The Significance of Airline Passenger Load factors. Washington, D.C.: Air Transport Association of America, 1980.
4. American Society of Heating, Refrigerating, and Air-Conditioning Engineers, Inc. ASHRAE Standard: Ventilation for Acceptable Air Quality. ASHRAE 62–1981. Atlanta, Ga.: American Society of Heating, Refrigerating, and Air-Conditioning Engineers, Inc., 1981.

5. Cain, W. S., B. P. Leaderer, R. Isseroff, L. G. Berglund, R. J. Huey, E. D. Lipsitt, and D. Perlman. Ventilation requirements in buildings, I: control of occupancy odor and tobacco smoke odor. Atmos. Environ. 17:1183–1197, 1983.
6. Engelhard Special Chemicals Division. Ozone Converter. Specification SIC-3. Service Information Letter SIL-3. Union, N.J.: Engelhard, 1984.
7. Halfpenny, P. F., and P. S. Starrett. Control of odor and irritation due to cigarette smoking aboard aircraft. ASHRAE J. 3(3):39–44, 1961.
8. Kerka, W. F., and C. M. Humphreys. Temperature and humidity effect on odor perception. ASHRAE Trans. 62:531–552, 1956.
9. Lorengo, D. E., and A. Porter. Aircraft Ventilation Systems Study: Final Report. DTFA-03-84-CO-0084. Atlantic City, N.J.: U.S. Federal Aviation Administration Technical Center, 1985. (draft)
10. Newman, W. H., and M. R. Viele. Engine Bleed Air Reduction in DC-10. NASA-CR-159846. Long Beach, Cal.: Douglas Aircraft Company, 1980.
11. Reese, J. Statement, pp. 51–81. In U.S. Senate, Committee on Commerce, Science, and Transportation, Subcommittee on Aviation (97th Congress, 2nd Session). Airliner Cabin Safety and Health Standards: Hearing on S. 1770, May 20, 1982. Serial No. 97–122. Washington, D.C.: U.S. Government Printing Office, 1982.
12. Society of Automotive Engineers. Aero-Space Applied Thermodynamics Manual. ARP-1168. New York, N.Y.: Society of Automotive Engineers, 1969.
13. Society of Automotive Engineers. Air-Conditioning Systems for Subsonic Airplanes. ARP-85E. Warrendale, Pa.: Society of Automotive Engineers, 1986.

14. Society of Automotive Engineers. Aircraft Cabin Pressurization Control Criteria. ARP-1270. Warrendale, Pa.: Society of Automotive Engineers, 1976.
15. Society of Automotive Engineers. Environmental Control System Contamination. AIR-1539. Warrendale, Pa.: Society of Automotive Engineers, 1981.
16. Thayer, W. W. Tobacco smoke dilution recommendations for comfortable ventilation. ASHRAE Trans. 88(2): 291–306, 1982.
17. U.S. Bureau of the Census. Statistical Abstract of the United States, 1985, p. 217. 105th ed. Washington, D.C.: U.S. Government Printing Office, 1984.
18. U.S. Federal Aviation Administration. Aircraft Air-Conditioning and Pressurization System Documentation. (unpublished compilation, 1985.)
19. U.S. Federal Aviation Administration. FAA Aviation Forecasts, Fiscal Years 1985–1996. FAA-APO-85-2. Washington, D.C.: U.S. Federal Aviation Administration, 1985.
20. U.S. Federal Aviation Administration. Transport Category Airplanes Cabin Ozone Concentrations. Advisory Circular 120-38. Washington, D.C.: U.S. Federal Aviation Administration, 1980.

3

Standards, Regulations, and Industry Practices

The Air Commerce Act of 1926[1] established for the first time federal responsibility for the regulation of civil aviation in the United States. The basic features of that act—registration, periodic examination, rating of aircraft as to airworthiness, and rating of the qualifications of crew members—were carried over into the Federal Aviation Act of 1958.[20] The Federal Aviation Act established the Federal Aviation Administration (FAA) as an independent agency responsible for regulating the safety of aviation and the Civil Aeronautics Board (CAB) as an independent agency responsible for its economic regulation. Neither act assigned direct responsibility for health effects associated with air travel. For example, the statutory basis for the regulation of smoking on aircraft by CAB is Section 404(a) of the Federal Aviation Act, which requires airlines to provide "safe and adequate service." FAA was absorbed into the Department of Transportation in 1966, and CAB was disbanded under the Airline Deregulation Act of 1978[2] and the Civil Aeronautics Board Sunset Act of 1984.[11] Many CAB responsibilities were transferred to the Office of the Secretary of Transportation. Today, the Secretary of Transportation is charged with responsibility for regulating air commerce so as to promote in the best way its development and safety in the United States and abroad by prescribing safety regulations and standards. But no federal office has direct responsibility for health effects associated with air travel. (For a historical description of relevant statutes, see Onstad and Roark.[31])

FAA's effort to satisfy this mandate is accomplished largely through the exercise of its regulatory powers in the promulgation of Federal Aviation Regulations (FARs) by headquarters in Washington, D.C., and the enforcement of FARs by regional offices. FARs are adopted in accordance with rule-making procedures that provide for public participation—from preliminary determination of need, through development and publication, to final promulgation and application (except in special circumstances, such as safety emergencies that require immediate action).

The processes by which FAA seeks to ensure the inherent safety and airworthiness of aircraft are type certification, which ensures that the design of particular new aircraft complies with statutes and applicable rules and regulations, and the establishment of standards that must be met by manufacturers and air carriers in the course of designing, producing, operating, and maintaining aircraft. Accordingly, FARs have been established that govern airworthiness standards for commercial transport airplanes and certification and operation of domestic, flag (foreign), and supplemental air carriers and commercial operators of large aircraft. (In keeping with the Committee's charge, regulations that govern noncommercial private aircraft and air taxi operators are not referred to here.)

An earlier National Research Council study[29] examined FARs and their implications for aircraft construction and maintenance. This chapter focuses on relevant sections of the FARs and their enforcement, especially with respect to their influence on the operating procedures of air carriers. Nevertheless, a few comments concerning type certification are appropriate.

FAA engineers cannot review each of the thousands of drawings, calculations, reports, and tests involved in type certification. But it must be certain that each design for a new aircraft meets all appropriate regulatory requirements. Thus, the system relies not only on the FAA staff, but also on the assistance rendered by aircraft company employees called designated engineering representatives (DERs), who review the design and design process to ensure, on behalf of FAA, compliance with all aspects of the appropriate regulations.

Once an aircraft has received type certification, the manufacturer may continue to produce aircraft according to the approved design as long as it wishes or until the type certification is amended by FAA. There is a strong incentive to produce according to the same design, in that substantial modifications must be submitted for certification. FARs often apply to designs certified after promulgation. Sometimes, however, a rule specifies that all aircraft type-certified after a particular date must be brought into compliance. For example, the requirement for escape-route markings near the floor applies to all aircraft certified after 1958 (virtually the entire commercial fleet).

Other rules—mostly those governing air carriers, as opposed to manufacturers—apply to all passenger-carrying aircraft, regardless of the date of type certification; an example is the requirement that smoke detectors and automatic fire extinguishers be installed in all lavatories. These generally specify a time limit for compliance.

The pattern of interaction between FAA and the aircraft manufacturers relies on mutual exchange and cooperation. The pattern of interaction with the carriers involves mainly continual surveillance and occasional sanctions, so it places more emphasis on inspection and enforcement than on review of design specifications and production. In 1979, a new set of regulations—Investigative and Enforcement Procedures of the Federal Aviation Regulations[25]—established a mechanism for filing formal complaints and prescribed enforcement procedures for issuance of orders of denial, cease-and-desist orders, and orders of compliance. These regulations include provisions for formal fact-finding under the Federal Aviation Act of 1958,[20] the Airport and Airway Development Act of 1970,[3] and the Hazardous Materials Transportation Act of 1974.[22]

This chapter reviews standards, regulations, and operating procedures with respect to several problems involving safety in the aircraft cabin. In particular, it examines regulations with respect to air quality itself, as well as regulations and guidelines that govern crew and passenger response to fire, depressurization, medical emergency, and ditching and evacuation. (Smoke and toxic fumes associated with cabin fires constitute one of the gravest hazards in aircraft emergencies, and rapid depressurization exposes passengers and crew members to hypoxia. Fires and depressurization directly involve aspects of cabin air quality and are addressed specifically in Chapter 4.)

The airlines, in conjunction with the manufacturers and the FAA, establish minima-equipment lists that define allowable operations for situations when equipment is inoperable. Compliance with this list dictates operations that will meet FAA regulations. Because relevant federal regulations necessarily leave considerable discretion to the air carrier in accommodating the different configurations of equipment on various aircraft, typical procedures of major North American air carriers with respect to these subjects are described here, as are similarities to and differences from foreign regulations. Most of the procedures and equipment relevant to safety and health in aircraft involve the behavior of passengers in some way, so the adequacy and efficacy of the provision of passenger safety information are also reviewed in this chapter.

U.S. REGULATIONS AND STANDARDS

The "Cabin Safety Subject Index" prepared by the FAA Civil Aeromedical Institute (CAMI) in January 1984 presents a long list of regulations and recommendations pertaining to safety standards and operating requirements of commercial aircraft.[33] This extensive index includes such items as the specification of who may be admitted to the cockpit, the width of aisles giving access to emergency exits, storage and use of galley equipment during takeoff and landing, use of public-address systems, and actions related to encounters with air turbulence. From this long list of safety provisions, three categories emerge as particularly important for this study: standards for cabin air quality, response to incidents and accidents (including fires, depressurization, and emergency landings), and other operating procedures (including those in medical emergencies).

Two parts of the FAR are relevant to the construction of aircraft and operation of commercial air carriers: Part 121, "Certification and Operations: Domestic, Flag, and Supplemental Air Carriers and Commercial Operators of Large Aircraft,"[10] and Part 25, "Airworthiness Standards: Transport Category Airplanes."[4] In addition, some Air Carrier Operations Bulletins (ACOBs), Advisory Circulars (ACs), and Airworthiness Directives (ADs) issued by FAA include relevant directions or recommendations for commercial air carrier operations. Many ACOBs have been collected in a consolidated reprint dated March 1985.[37]

Ventilation

The airworthiness standards require that cockpit and cabin air be free of harmful or hazardous concentrations of gases or vapors.[47] They specify that carbon monoxide concentrations must be less than 1 part in 20,000 parts of air (50 ppm) and that carbon dioxide concentrations must not exceed 3% by volume (sea level equivalent), or 30,000 ppm. The carbon monoxide and carbon dioxide standards were incorporated into the Federal Aviation Act of 1958. The carbon monoxide standard apparently originated in requirements related to exhausts from internal-combustion heaters.[35] The carbon dioxide standard first appeared as an amendment to the Civil Air Regulations in 1952.[36] There are no requirements for monitoring of carbon monoxide or carbon dioxide in the cockpit or cabin during flight. Nor are there explicit requirements concerning ventilation rates for passenger cabins; the regulations state only that "each passenger or crew compartment must be suitably ventilated".[48]

Ozone

Ozone contamination of aircraft cabins is a problem during high-altitude or high-latitude flights. Ozone is a known irritant and has been associated with some health effects. The FARs specify that cabin ozone concentration must not exceed 0.1 ppm by volume sea level equivalent (SLE) time-weighted average during any 3-h interval, nor exceed 0.25 ppm (SLE) at any time.[9] This standard is also found in the regulations governing air carriers, which state that the ozone concentration requirement may be satisfied either by air treatment equipment (usually a catalytic converter) that maintains cabin ozone concentrations at or below this requirement or through appropriate scheduling of flight plans on the basis of average atmospheric ozone.[9] This rule was promulgated in 1980 on the basis of extensive review of data concerning human and animal exposure to ozone and with opportunity for public input to the rule-making process.[39] There is no FAA requirement for in-flight monitoring of ozone (the United Kingdom requires monitoring on flights above 49,000 ft). Compliance with the federal regulation is based on performance of the air treatment equipment at the time of installation or on the flying of routes and altitudes that avoid high ozone concentrations.

Fires

Onboard fire threatens passengers not only directly, because of the possibility of burn injury and inhalation of smoke and toxic fumes, but also indirectly, because of the possibility of damage to the structural integrity of the aircraft and its ability to remain in controlled flight and of increased difficulty of escape once the aircraft has landed and stopped. Much can be done to reduce the ease of ignition, inhibit the propagation of flame, and reduce smoke and toxic fumes by careful selection and use of materials. These issues are addressed separately in Chapter 4. The procedures and equipment described in Appendix C deal with firefighting by the crew and with passenger behavior in fire and other emergencies.

Of the regulations and recommendations referred to in this chapter, those concerning fires are the most extensive (see Table C-1). That was true even before the recent promulgation of rules and recommendations that followed the cabin fires near Cincinnati, Ohio on June 2, 1983, and at Tampa International Airport on June 25, 1983. Additional regulations have been proposed or implemented since those events.

Neither the regulations nor the recommendations specify emergency procedures in case of fire. A few recommendations are made—for example, to review emergency procedures concerning operation of lower-lobe galleys (below the main floor) in jumbo jets. Current regulations specify some nonemergency procedures, such as prohibition of smoking while the "no smoking" sign is lighted[32] or provisions for maintenance of fire-susceptible areas.[38] There is extensive prescription of safety equipment with respect to fires, including provisions for the prevention of fires ("no smoking" signs in lavatories[14]), for detection of fires (smoke detectors in lavatories and galleys[38]), for extinguishing of fires (automatic and hand fire extinguishers[38]), for protective breathing equipment and firefighting training for crew,[44] and for passenger escape from smoke-filled cabins (floor-proximity escape-route markings, i.e., exit routes visible when there are no sources of light more than 4 ft above the floor[42]).

Crew training is to include not only initial familiarization with equipment and procedures, but periodic "hands-on" refreshers. Passengers are to be briefed by announcements concerning smoking in the cabin and in lavatories.[8]

Depressurization

Sudden depressurization of the aircraft cabin threatens passengers with hypoxia. The regulations and standards for pressurization and depressurization are somewhat less extensive than those for fires (see Table C-2). Emergency procedures are not specified, but minimal flows of supplemental oxygen are specified in terms of equipment, altitude, duration at altitude, and other factors.

Equipment must be available to deliver supplemental oxygen for crew and passengers whenever the airplane is operated at an altitude of over 10,000 ft.[18] For flight above 25,000 ft, an automatic system to deploy supplemental oxygen equipment in the event of sudden depressurization is required; and portable oxygen equipment with a 15-min supply must be provided for cabin crew.

Proper use of continuous-flow passenger masks has proved to be a satisfactory intermediate measure for countering inadequate supply of oxygen for cabin altitudes of up to 40,000 ft. Most of the problems associated with these masks appear to be related to the lack of timely or proper use—activating the oxygen flow, covering both nose and mouth, and ensuring a tight fit. Because of their higher degree of physical activity, cabin crew have only 15–20 s to don and activate masks before adverse effects set in at 40,000 ft; passengers have about 40 s.[23]

Crew training includes both initial familiarization and periodic refreshers with respect to supplemental oxygen equipment and familiarization with medical symptoms associated with hypoxia and depressurization.

Before a flight above 25,000 ft, passengers must be briefed on the use of supplemental oxygen equipment.

Medical Emergencies

A recent court decision reversed an FAA decision that it did not have authority to make air carriers supply their aircraft with medicine and emergency medical equipment to treat general health emergencies [6] and held that FAA can proceed with rule-making if it deems such action to be advisable. In response, a rule has been adopted that requires much more extensive medical kits than had been required. [40] This new rule would considerably extend the scope of FAA regulations with respect to medical emergencies.

No regulation of emergency procedures is specified (see Table C-3). With respect to nonemergency procedures, conditions are determined under which a passenger may carry and operate oxygen equipment for medical reasons.

The content and number of first-aid kits for injuries likely to occur in flight or in minor accidents are specified. The new regulation[40] adds the requirement of a medical kit containing equipment and drugs required for life support during medical emergencies (including myocardial infarction, severe allergic reactions, acute asthma, insulin shock, protracted seizures, and childbirth).

Crew training includes instruction in emergency procedures and familiarization with first-aid equipment and practices. The regulation requires familiarization with the medical kit, although this does not include training in all the medical procedures possible with the equipment in the kit. It part, the purpose of the kit is to enable passengers with appropriate medical training to respond to medical emergencies, as well as to extend the crew's capability to provide advanced first-aid techniques.

No special provisions are made for passenger briefings concerning medical emergencies.

Ditching and Evacuation

Because of the extensive requirements for appropriate design of the airframe and ancillary equipment, ditching and evacuation are subject to extensive regulations and recommendations (see Table C-4). Recommendations for emergency procedures include suggestions for preparation of passengers for an emergency landing. Procedural requirements include assurance that the crew are fully familiar with operation of emergency equipment and evaluation of proper bracing positions with due regard to seat spacing.

Each passenger-carrying landplane emergency exit (except over-the-wing exits) that is more than 6 ft from the ground when the landing gear is extended must have equipment for assisting occupants in descending. [17] On all flights that include extended over-water operation, flotation devices must be within easy reach of all passengers, and liferafts must be sufficient to accommodate all occupants.[15]

Crew training includes instruction and periodic refreshers in emergency procedures and equipment for each type of airplane. Passenger briefings are to include the location of emergency exits and, in over-water flights, the location of flotation devices and methods of donning and inflating life preservers.

Additional Passenger Briefings

Before takeoff, passengers are to be briefed concerning smoking, location of emergency exits, use of safety belts (including how they are fastened and unfastened), and location and use of required emergency flotation devices (see Table C-5). After takeoff, an announcement must be made, immediately before or after the seatbelt sign is turned off, that passengers should keep their seatbelts fastened while seated.[8]

FOREIGN REGULATIONS

Many foreign aviation standards and regulations draw heavily on the U.S. FARs, but there is also considerable activity in the field of air safety elsewhere. We will not attempt to include an exhaustive review of foreign standards and regulations here. However, we will review the European Civil Aviation Conference (ECAC) regulations to illustrate typical similarities and differences between U.S. and foreign regulations. For example, ECAC develops, for application by its member states as their own national regulations, "uniform requirements for the following emergency and safety airborne equipment for large aircraft: a) emergency oxygen equipment, b) evacuation equipment, c) sea rescue and survival equipment, and d) possible crash or fire survival equipment."[19] The ECAC Working Group on Cabin Safety recently approved amendments to the ECAC regulations, but these were not available at the time of the Committee's deliberations.

Most ECAC requirements are based on the U.S. FARs. However, there are differences.[19] With respect to fires, the ECAC regulations prohibit the use of dry chemical extinguishers in the cockpit or in any compartment not separated by a partition from the cockpit. No attention is given to medical emergencies in ECAC requirements, nor are standards recommended for carbon monoxide, carbon dioxide, or ozone. However, regulations in the United Kingdom require continuous monitoring of ozone on all flights above 49,000 ft. (In practice, this applies only to the Concorde and a few corporate jets.)

Some ECAC requirements with respect to ditching and evacuation are different. An explicit formula is provided for reducing the number of passengers and seating distributions if an exit becomes inoperative at an airport where it is not practical for it to be repaired or replaced. Seat cushions are not considered to be acceptable flotation devices.

There are also differences as to passenger briefing and crew training and responsibilities. For example, the oral briefing on the location of emergency exits may be omitted if the subject is covered by oral reference to the briefing cards. The number of airplane types in which cabin crew are qualified at any particular time is to be limited, and the crew is to receive training on survival at sea, on uninhabited terrain, and in extreme climatic conditions.

INDUSTRY OPERATING PROCEDURES

Airline operators are allowed considerable discretion in complying with safety regulations. That is largely inevitable, given the different configurations of airframe and equipment, ranging, for example, from large wide-body airplanes to smaller narrow aircraft. Most of the relevant FARs apply to airplanes with capacity for 20 or more passengers. Thus, to some degree, cabin safety depends on the standard operating procedures of the individual carriers for the different types of airplanes.

The Committee attempted to elicit descriptions of operating procedures from U.S. and foreign flag air carriers. However, the response was not sufficient to be representative of the industry as a whole. It does appear, however, that ECU packs and other ventilation equipment are generally to be fully activated in case of smoke or fire. Table C-6 describes the firefighting procedures and training of one foreign carrier.

From these examples, two issues emerge: the appropriateness of FAA standards and regulations and the degree to which the industry operating procedures satisfy those standards and regulations. FAA regulations were discussed earlier in this chapter. The determination of whether the carriers' procedures comply is the responsibility of FAA inspectors, who regularly inspect various aspects of carrier operations, as described in the next section.

FAA INVESTIGATION AND ENFORCEMENT

FAA has the power and duty to investigate reported violations and to determine whether enforcement action is warranted. Any person who knows of a violation is encouraged to report it to appropriate personnel of any FAA regional or district office.[34] Air carriers are required to report accidents and incidents, as well as other interruptions to service. FAA is also empowered to initiate an investigation at any time with respect to any matter within its jurisdiction.

Enforcement alternatives provided by the statutes and regulations include amendment, modification, suspension, or revocation of certificate if aviation safety and the public interest require it. Civil penalties of up to $1,000 for each violation (up to $10,000 for a violation associated with a hazardous material) can be imposed. There are several informal methods to seek relief from these enforcement processes; if these fail, more formal proceedings take place before FAA hearing officers, the National Transportation Safety Board (NTSB), or a U.S. district court.[31]

Investigation of Violations

FAA relies primarily on the informal investigative process to obtain information.[31] Most complaints received by FAA are informal and are acted on through informal investigation. If a formal complaint is received, the FAA administrator must determine within an allotted time whether there are reasonable grounds for investigating it.[21] If reasonable grounds exist, an informal investigation is initiated or an order of investigation issued.

Formal fact-finding investigations may be initiated by FAA counsel whenever it is determined that informal procedures are inappropriate or inadequate. The sole purpose of formal proceedings is to determine whether the available facts warrant further action. If the evidence warrants no further action, the investigative file is closed. If the investigating office determines that enforcement action is justified, appropriate administrative or legal action is taken.

Enforcement Actions

The FAA enforcement program consists of administrative and legal enforcement actions.[31] Administrative actions are analogous to a warning ticket, and there is no right of appeal. In legal actions, FAA has the burden of proof; in contrast, the applicant has the burden of proof in cases of discretionary denial of an application for a license or certificate on medical grounds. The nature of the violation and its impact on aviation safety determine the type of enforcement action taken.

Administrative enforcement provides FAA field inspectors a means of dealing with minor violations and is used when there is no major unsafe condition and when a violation was inadvertent. In this situation, an alleged violator must have a "constructive attitude" toward complying with the regulations and must not have been involved in previous similar violations. If corrective action is taken, the violator avoids further enforcement action. Failure of the violator to correct the problem can result in civil penalties or initiation of other legal enforcement proceedings.

Legal action usually begins with certificate action, such as suspension or revocation, or with imposition of a civil penalty.[27] Other actions intended to achieve compliance are aircraft seizure, issuance of orders (such as cease-and-desist orders), injunctive relief, and imposition of criminal penalties. Suspension of a certificate usually occurs in a case of serious operational violation and when a possible lack of qualification could be corrected by remedial action or retraining; suspension can also be used for disciplinary purposes, if the nature of the violation warrants it.

Civil penalties can be imposed if the facts and circumstances are too serious to be handled by administrative action and either the violator does not hold a current certificate or there is no question as to the qualification of the certificate holder.[12] [13] In most cases, violations involve questions of whether the certificate holder has carried out its obligations under the regulations, rather than questions of qualification as a certificate holder. The maximal penalty is $1,000 for each violation. If a violation is continuing, each operation constitutes a separate violation. In determination of the amount of the penalty, the nature, circumstances, extent, and gravity of the violation and the ability of the violator to absorb the sanction are taken into account. A civil penalty usually is not imposed if there has been a certificate action or criminal penalty for the same violation.

If enforcement actions fail to deter repeated violations, a complaint can be filed in the appropriate U.S. district court requesting issuance of an injunction or other process to restrain the violator from further action. In emergency situations, issuance of such an order is preferred to all other enforcement actions.

Review and Appeal

If the FAA administrator denies issuance or renewal of a crewman's certificate, the applicant can petition for the action to be reviewed by NTSB within 60 d.[31] The burden of proof in denial of a medical certificate lies with the applicant. If a certificate is denied or revoked by the FAA administrator, an appeal can be made to NTSB within 20 d. In these actions, the administrator must show that the certificate was suspended or revoked justifiably.

On petition or appeal, an NTSB administrative law judge conducts a formal proceeding under the NTSB rules of practice and makes initial findings of fact and conclusions of law. Ultimately, all enforcement actions by FAA and NTSB are reviewable by U.S. courts of appeals.

Effectiveness and Adequacy of FAA Inspection and Enforcement

The Committee has not been able to locate scientific, peer-reviewed studies of the implementation of the provisions for inspection and enforcement described above on which to base an evaluation of their effectiveness and adequacy. However, the General Accounting Office (GAO) has conducted an analysis of FAA inspection of a sample of commercial air carriers during 1984.[46] The results of that analysis are the basis of the following discussion.

GAO studied the type, frequency, and results of FAA inspections. Five of FAA's nine regions were selected; they accounted for responsibility for inspection of about 70% of the nation's approximately 500 scheduled Part 121 and Part 135 air carriers. At each district office, air carriers were randomly selected; 40 of 73 Part 121 carriers and 52 of 112 scheduled Part 135 carriers were selected. All FAA inspections of the 92 selected carriers in 1984, whether conducted by those offices or by other FAA units, were reviewed—about 12,000 inspection reports. The study did not examine FAA activities concerning certification or investigations of accidents and incidents.

Data were grouped into FAA's three inspection categories: operations, maintenance, and avionics. On the basis of principles of descriptive statistics, GAO calculated that the data in its charts were at the 95% level of confidence. The GAO study distinguished inspections before the National Air Transportation Inspection (NATI) program, which lasted about 90 days, and those during and after it. Table 3-1 presents the percentages of inspections that were considered satisfactory and unsatisfactory and that were not classified.

Concern has been expressed that commercial safety regulations are avoided by some aircraft operations. For example, one GAO report focused on the use of leased private aircraft (subject to less stringent Part 91 regulations) in situations that qualify as commercial operations.[45]

In addition, although all in-flight fires must be reported to NTSB and FAA, the NTSB and FAA data bases concerning fires do not correspond exactly, because the FAA data also include incidents found only in the mechanical reliability reports. These differences might result from differing interpretations of the severity of a fire. Furthermore, a paper presented to a major conference on cabin safety in 1984 pointed out that United Airlines reported that it had about 60 fires in 1980–1984, but NTSB data contain only 40 fires over the 10 years before 1984.[24] This suggests that many incidents never enter the aviation safety reporting system.

TABLE 3-1 Results of FY 1984 FAA Inspections of 92 Sampled Air Carriers in Five FAA Regionsa

	Pre-NATI,[b] 10/1/83 – 2/29/84			During and After NATI,[b] 3/1/84 – 9/30/84		
Category/ FAA Region	% Sat.	% Unsat.	% Unmarked	% Sat.	% Unsat.	% Unmarked
Operations:						
Alaska	90	10	0	85	12	3
Eastern	99	1	0	95	5	0
Northwest Mt.	92	7	1	91	8	1
Southern	92	7	1	88	12	0
Western Pac.	96	4	1	92	7	1
Maintenance:						
Alaska	96	3	1	92	7	0
Eastern	62	22	16	73	26	1
Northwest Mt.	97	2	2	92	7	1
Southern	84	13	3	82	17	1
Western Pac.	87	11	2	88	11	1
Avionics:						
Alaska	100	0	0	95	2	3
Eastern	92	2	5	88	9	3
Northwest Mt.	94	4	2	93	6	1
Southern	87	10	3	87	9	3
Western Pac.	90	7	3	92	7	1

[a] Data from U.S. General Accounting Office.[46]

[b] National Air Transportation Inspection program, conducted by FAA from about March 1, 1984, to June 1, 1984.

On balance, it is not possible for the Committee to determine the facts concerning the effectiveness and adequacy of FAA enforcement and inspection.

ADEQUACY AND EFFICACY OF PASSENGER SAFETY INFORMATION

The issue of the adequacy and efficacy of passenger safety briefings presents a genuine dilemma. There is considerable variation in the articulation and presentation of safety briefings by flight crews. Furthermore, many in the flying public are rather familiar with passenger briefings and, believing that they already know the content, tend to ignore them. Yet statistics reflecting passenger response in emergency situations suggest that most passengers either do not understand the instructions or do not apply their knowledge when confronted with an emergency.[30] Although this situation has been studied, there is not a sufficient scientific basis for specific policy recommendations. Empirical research results do, however, indicate some of the principal factors involved and suggest experimental approaches that might be pursued to develop a better understanding of the problems involved.

The basic problem—provision of routine safety information that must be recalled under emergency conditions—involves phenomena related to a variety of research subjects, including interest and attention span, comprehension and retention, and recall under stress. A review of the vast literature on these and related topics is beyond the scope of this study, so we focus here on topics that appear to be most relevant to commercial air travel.

The efficacy of alternative onboard presentations (verbal announcements, videotape presentations, and safety cards and placards) is not addressed here. A recent study by NTSB addressed such comparisons in detail; [30] it drew from a series of theoretically unrelated empirical studies of immediate relevance to the conditions of alternative onboard announcements.

Factors Influencing Passenger Emergency Behavior

NTSB studies have focused on "maladaptive passenger behavior" in emergencies, ranging from inability to perform such emergency tasks as donning a lifevest to total inaction. The results suggested several factors that lead to such behavior: the inappropriateness or inaccuracy of information given to passengers, passenger indifference to safety information, the belief apparently held by some passengers that they are immune to injury, and the common belief that airplane accidents are not survivable and that passengers consequently have no influence on whether they will survive an accident.[30]

During the 1970s, McDonnell Douglas studied passenger behavior in actual emergency situations and proposed three methods to stimulate improvement: learning by trial and error, training or instruction, and clear and forceful instructions and action by the crew. McDonnell Douglas concluded that the combination of forceful cabin crew leadership with provision of passenger instruction is more effective than either alone.[30] That is an important finding, especially because the study examined the specific situation of passenger briefings and response in the setting of commercial aircraft. But the finding is limited, because it is not related to a more general model of information communication, comprehension, and recall.

Attention, Comprehension, and Recall under Stress

Industry observers have often suggested that a mild degree of anxiety increases the attention given to safety briefings and instruction cards. Too much anxiety and fear, however, often result in a "disaster syndrome" in which psychologic blocking results in inaction or inappropriate response. Berkun and others presented briefing-card information to three groups of subjects.[7] Those in group 1 took off, were informed during the flight that they would have to perform an emergency ditching operation, and were then tested in flight on the briefing-card information. Those in group 2 took off, were not informed of an impending emergency, and were tested in flight on the same information. Subjects in group 3 remained on the ground, but were also tested. Time between presentation and testing was the same for all groups. Results showed that group 1 performed significantly worse than group 2, and group 2 significantly worse than group 3.

Those results correspond to a model of the effect of emotion on the use of information—a model that relies on extensive animal and human experimental results.[16] The model predicts that, in heightened emotional states (in this case anxiety), a subject reduces the amount of attention given to "peripheral" cues and focuses on cues of most central importance. In a mild state of anxiety, for example, a person might ignore other activities in the cabin and attend to the safety announcement; this ought to result in an increase in retention and response. With greater anxiety, such as the prospect of ditching, a greater number of peripheral mental cues are ignored, perhaps even including some that are relevant for appropriate response. In other words, increasing anxiety eventually results in degeneration of response. According to this model, a course of action is facilitated or disrupted by emotion, depending on the complexity of the action and on the cues attended to in the particular emotional state.

The notion of facilitation or disruption of attention, comprehension, recall, and action by heightened emotion of various degrees appears to apply to many of the studies that have been conducted in the airline industry. For example, results of a McDonnell Douglas study suggested that explicit emergency evacuation instructions in a preflight briefing would not cause anxiety, but rather would reduce anxiety if properly presented.[5] However, interviews of survivors of aircraft accidents suggested that, of the four common responses to the extreme stress of emergency situations—strengthening of resources, attacking the threat, avoiding it, or remaining inactive—the most common response by passengers during aircraft emergencies was inaction.[26]

General psychologic research has suggested that under such conditions of high stress greater proficiency is maintained for the simplest perceptual-motor tasks. But the airline passenger must perform rather complex tasks in emergency situations—correctly donning a life preserver while seated with the seatbelt fastened and donning an oxygen mask, activating the oxygen flow, and ensuring a tight fit—and perform them properly on the first try under conditions of extreme emotional stress. The aim of passenger safety briefings is not subject to simple resolution. Presentation of information under routine conditions in such a way as to ensure recall under extreme emotional stress is a difficult task.

Improving Passenger Safety Briefings

Although the results of studies noted above are insufficient to provide specific recommendations about how to increase the adequacy and efficacy of the provision of passenger safety information, one possible improvement is apparent. Under conditions of stress, a person is more likely to be able to perform perceptual-motor operations that have been well learned. Therefore, it would probably be most efficacious to provide passengers an opportunity to learn how to don masks and life preservers. The advisability of providing passengers a greater opportunity to familiarize themselves with the opening of escape windows and doors is less clear. In at least one instance, passengers have opened emergency escape doors when there was an engine fire and the crew judged emergency evacuation of the aircraft to be unnecessary.

Hands-on training might be impracticable, because of the variations in equipment among various aircraft. A cost-effective alternative might be to provide detailed video presentations of safety procedures in waiting lounges. It is well within the capability of current technology to store presentations describing the various aircraft used by a given airline, or even several airlines, to be viewed before passengers board a particular aircraft. Such safety information might be presented in a number of different formats. For example, it might be presented "on demand," so as to obtain a measure of the flying public's awareness and interest, or it might be presented in conjunction with simple learning games, which could be used both to reinforce the information and to measure its understandability.

That approach would be aimed at increasing the overall understanding of safety procedures in the flying public. The research results do not suggest an advantage associated with the presentation of all relevant information to every passenger on every flight. However, it is obvious that all essential information should be available to any passenger who wants it. Video presentations in waiting lounges might have the double advantage of presenting relevant information in considerably greater detail than is commonly the case today and presenting it in a manner that is flexible enough to serve the needs of both frequent fliers and neophytes.

Again, the basic elements of the problem are apparent: the difficulty in attracting and keeping passenger attention, the difficulty in communicating complicated perceptual-motor procedures, and the latent difficulty in recalling this information under conditions of extreme stress. The suggested approach does not remove any of these difficulties, but does present the possibility of developing incremental improvements that would permit gradual increases in theoretical and empirical understanding of the complex phenomena involved.

OVERVIEW

Cabin safety procedures, equipment, and passenger information have all received considerable attention and are the subject of many federal regulations and guidelines. But the regulations necessarily leave considerable discretion to individual air carriers, because of variations in the configuration of the equipment in aircraft. Therefore, in accordance with the regulations, airlines have developed equipment and procedures that they feel are appropriate for routine conditions and emergencies in air travel. This chapter has reviewed standards, regulations, and procedures that have been developed by FAA and the industry.

A discussion of cabin safety would be incomplete without reference to a number of recent actions of FAA. In particular, FAA has proposed a rule concerning the use of specific materials in cargo and baggage compartments;[41] promulgated a final rule requiring emergency escape-path markings that are visible when all sources of cabin lighting more than 4 ft above the floor are obscured by smoke;[42] promulgated a final rule requiring air carriers to provide medical kits containing equipment and drugs for use in the treatment of injuries or medical conditions that occur during flight;[40] proposed a rule on fire protection requirements for cargo or baggage compartments that includes provision for at least two Halon extinguishers and for inspections and repairs of lavatory electric components on some aircraft;[38] proposed a rule establishing new fire test criteria for type certification of aircraft, which requires cabin interiors to correspond with the criteria, including the retrofitting of aircraft constructed since 1958;[43] and proposed a rule governing the availability and performance of breathing devices to protect the crew from smoke and toxic fumes.[44] In addition to these regulatory actions, FAA is conducting research in related subjects.[28]

It appears advisable to review the advantages and disadvantages of a carefully designed program of passenger information aimed at developing a better understanding of passenger response to safety instructions. Consideration should be given to conducting quizzes during flight to see, for example, what proportion of passengers have retained the key features of the safety briefings. Empirical research on the specific conditions of air travel can be combined with more general psychological evidence in ways that suggest an approach that ought to reveal useful scientific data while permitting incremental improvements over current practice. Although the suggestions presented in this chapter are motivated principally by a desire to improve passenger recall under the heightened stress of emergencies, they should also permit the collection of data that would illuminate the issues of attracting and keeping passenger attention and of comprehension of safety material presented.

REFERENCES

1. Air Commerce Act of 1926. PL 69–254. U.S. Statutes at Large 44:568–576, 1926.
2. Airline Deregulation Act of 1978. PL95-504 U.S. Statutes at Large 92:1705–1755, 1978.
3. Airport and Airway Development Act of 1970. PL 91-258. U.S. Statutes at Large 84:219–253, 1970.
4. Airworthiness standards: Transport category airplanes. Code of Federal Regulations, Title 14, Pt. 25. Washington, D.C.: U.S. Government Printing Office, 1985.
5. Altman, H. B., D. A. Johnson, and D. I. Blom. Passenger emergency evacuation briefing cards: Recommendations for presentation style (summary report). Vol. 2, pp. 455–474. In Survival and Flight Equipment Association. Proceedings of the Eighth Annual Symposium, Las Vegas, Nev., Sept. 28, 1970. Van Nuys, Cal.: Survival and Flight Equipment Association, 1970.
6. Bargmann, Eve, et al., Petitioners, v. J. Lynn Helms, Administrator, Federal Aviation Administration, Drew Lewis, Secretary of Transportation, Department of Transportation, Respondents. 715 F. 2d. 638. U.S. Court of Appeals D.C. Circuit 230:164–169, 1983.

7. Berkun, M. M., H. M. Bialek, R. P. Kern, and K. Yagi. Experimental studies of psychological stress in man. Psychol. Monogr. Gen. Appl. 76(15):1–39, 1962.
8. Briefing passengers before takeoff. Code of Federal Regulations, Title 14, Pt. 121.571. Washington, D.C.: U.S. Government Printing Office, 1985.
9. Cabin ozone concentrations. Code of Federal Regulations, Title 14, Pt. 25.832. Washington, D.C.: U.S. Government Printing Office, 1985.
10. Certification and operations: Domestic, flag and supplemental air carriers and commercial operators of large aircraft. Code of Federal Regulations, Title 14, Pt. 121. Washington, D.C.: U.S. Government Printing Office, 1985.
11. Civil Aeronautics Board Sunset Act of 1984. PL98-443. U.S. Statutes at Large 98:1703–1713, 1984.
12. Civil penalties: Federal Aviation Act of 1958. Code of Federal Regulations, Title 14, Pt. 13.15. Washington, D.C.: U.S. Government Printing Office, 1985.
13. Civil penalties: Hazardous Materials Transportation Act. Code of Federal Regulations, Title 14, Pt. 13.16. Washington, D.C.: U.S. Government Printing Office , 1985.
14. Compartment interiors. Code of Federal Regulation Title 14, Pt. 25.853. Washington, D.C.: U.S. Government Printing Office, 1985.
15. Ditching equipment. Code of Federal Regulations, Title 14, Pt. 25.1415. Washington, D.C.: U.S. Government Printing Office, 1985.
16. Easterbrook, J. A. The effect of emotion on cue utilization and the organization of behavior. Psychol. Rev. 66:183–201, 1959.

17. Emergency exit arrangement. Code of Federal Regulations, Title 14, Pt. 25.809. Washington, D.C.: U.S. Government Printing Office, 1985.
18. Equipment standards for oxygen dispensing units. Code of Federal Regulations, Title 14, Pt. 25.1447. Washington, D.C.: U.S. Government Printing Office, 1985.
19. European Civil Aviation Conference. Joint Requirements for Emergency and Safety Airborne Equipments. ECAC.CEAC Doc. No. 18. Paris: European Civil Aviation Conference, 1979.
20. Federal Aviation Act of 1958. PL 85–726. U.S. Statutes at Large 72:731–811, 1958.
21. Formal complaints. Code of Federal Regulations, Title 14, Pt. 13.5. Washington, D.C.: U.S. Government Printing Office, 1985.
22. Hazardous Materials Transportation Act. PL 93–633. U.S. Statutes at Large 88:2156–2173, 1974.
23. Higgins, E.A. Protective breathing: Oxygen mask use/problems, pp. 61–65. In Flight Safety Foundation, Inc. Proceedings of Cabin Safety Conference and Workshop, December 11–14, 1984. Washington, D.C.: U.S. Federal Aviation Administration, Office of Aviation Safety, 1985.
24. Inflight Occupant Protection Working Group. Inflight Occupant Protection Workshop: Summary of proceedings, pp. 167–171. In Flight Safety Foundation, Inc. Proceedings of Cabin Safety Conference and Workshop, December 11–14, 1984. DOT/FAA/ASF100–85/01. Washington, D.C.: U.S. Federal Aviation Administration, Office of Aviation Safety, 1985.
25. Investigative and enforcement procedures. Code of Federal Regulations, Title 14, Pt. 13. Washington, D.C.: U.S. Government Printing Office, 1985.

26. Johnson, D. A. Behavioral inaction under stress conditions similar to the survivable aircraft accident, pp. 42–47. In Survival and Flight Equipment Association. Proceedings of the Ninth Annual Symposium, Las Vegas, Nev., Sept. 27, 1971. Canoga Park, Cal.: Survival and Flight Equipment Association, 1972.
27. Legal enforcement actions. Code of Federal Regulations, Title 14, Pt. 13, Subpart C. Washington, D.C.: U.S. Government Printing Office, 1985.
28. McSweeny, T. E. FAA cabin safety activities, pp. 30–44. In Flight Safety Foundation, Inc. Proceedings of Cabin Safety Conference and Workshop, December 11–14, 1984. Washington, D.C.: U.S. Federal Aviation Administration, Office of Aviation Safety, 1985.
29. National Research Council, Committee on FAA Airworthiness Certification Procedures. Improving Aircraft Safety: FAA Certification of Commercial Passenger Aircraft. Washington, D.C.: National Academy of Sciences, 1980.
30. National Transportation Safety Board. Safety Study, Airline Passenger Safety Education: A Review of Methods Used to Present Safety Information. Washington, D.C.: National Transportation Safety Board, 1985. (draft)
31. Onstad, C. H., and D. R. Roark, Jr. Air safety: Enforcement of the Federal Aviation Regulations, pp. 195–222. In A. Kean, Ed. Essays in Air Law. The Hague: Martinus Nijhoff Publishers, 1982.
32. Passenger information. Code of Federal Regulations, Title 14, Pt. 121.317. Washington, D.C.: U.S. Government Printing Office, 1985.
33. Pollard, D. W., J. A. Steen, W. J. Biron, and R. L. Cremer. Cabin Safety Subject Index. FAA-AM-84-1. Oklahoma City, Okla.: U.S. Federal Aviation Administration, Civil Aeromedical Institute, 1984.

34. Reports of violations. Code of Federal Regulations, Title 14, Pt. 13.1. Washington, D.C.: U.S. Government Printing Office, 1985.
35. Society of Automotive Engineers, Inc. Aerospace Standard: Heaters, Aircraft, Internal Combustion Heat Exchanger Type. AS 143. New York, N.Y.: Society of Automotive Engineers, Inc., 1943.
36. U.S. Civil Aeronautics Board. Ventilation: Final rule. Federal Register 17(5 Feb.):1094, 1952.
37. U.S. Federal Aviation Administration. Air Carrier Operations Bulletins Consolidated Reprint. Washington, D.C.: U.S. Federal Aviation Administration, 1985.
38. U.S. Federal Aviation Administration. Airplane cabin fire protection: Final rule. Federal Register 50(29 March):12726–12733, 1985.
39. U.S. Federal Aviation Administration. Airplane cabin ozone contamination: Final rule . Federal Register 45(21 Jan.):3880–3883, 1980.
40. U.S. Federal Aviation Administration. Emergency medical equipment: Final rule. Federal Register 51(9 Jan.):1218–1223, 1986.
41. U.S. Federal Aviation Administration. Fire protection requirements for cargo or baggage compartments: notice of proposed rulemaking. Federal Register 49(8 Aug.):31830–31838, 1984.
42. U.S. Federal Aviation Administration. Floor proximity emergency escape path marking: Final rule. Federal Register 49(26 Oct.):43182–43186, 1984.
43. U.S. Federal Aviation Administration. Improved flammability standards for materials used in the interior of transport category airplane cabins: notice of proposed rulemaking. Federal Register 50(16 April):15038–15053, 1985.

44. U.S. Federal Aviation Administration. Protective breathing equipment: notice of proposed rulemaking. Federal Register 50(10 Oct.):41452–41459, 1985.
45. U.S. General Accounting Office. Commercial Safety Regulations Are Avoided by Some Large Aircraft Operators. CED-79-10. Washington, D.C.: U.S. General Accounting Office, 1978.
46. U.S. General Accounting Office. Compilation and Analysis of the Federal Aviation Administration's Inspection of a Sample of Commercial Air Carriers. GAO/RCED-85-157. Washington, D.C.: U.S. General Accounting Office, 1985.
47. Ventilation. Code of Federal Regulations, Title 14, Pt. 25.831. Washington, D.C.: U.S. Government Printing Office, 1985.
48. Ventilation. Code of Federal Regulations, Title 14, Pt. 121.219. Washington, D.C.: U.S. Government Printing Office, 1985.

4

Air Quality in Emergency Situations

Chapter 2 describes the physical factors that influence airliner cabin air quality under normal operating conditions, and Chapter 3 describes the federal regulations and industry operating procedures that bear on air quality. This chapter focuses on the effects of emergency situations on cabin air quality.

Two in-flight emergency situations affect cabin air quality: fire and cabin depressurization. Not only can fire lead to deterioration of the structural integrity of the aircraft and its ability to remain in controlled flight, but the resulting smoke and toxic combustion products and ultimately the fire itself constitute direct hazards to passengers and crew. The main threat to passengers in sudden depressurization is hypoxia.

ONBOARD FIRES

Providing protection from fire in airliners is a complicated matter. Cabin interiors are furnished and lined with potentially flammable materials, and passengers are tightly packed in a relatively small, confined enclosure. Inaccessible compartments contain potential ignition sources and combustible materials, and wing tanks carry thousands of gallons of highly flammable aviation fuel. Given these conditions, the average of 32 deaths a year from the effects of fire involving U.S. air carriers between 1965 and 1979 might seem remarkably low, compared with the figures on other modes of transport.[14] But it is a major concern. An estimated 15% of all deaths in domestic air carrier accidents during that period have been attributed to the effects of fire.[15]

An analysis of air transport accidents in North Atlantic Treaty Organization (NATO) countries between 1964 and 1975 revealed (when such information could be determined) that injuries and deaths were due primarily to the postcrash effects of fire, smoke, and toxic fumes and only secondarily to crash impact itself.[17] The aircraft used in NATO countries are largely of American manufacture and meet American standards, so data on accidents in these countries should be considered with the American data; that increases the apparent incidence of fire-related death.

Three accidents have played an especially prominent role in increasing awareness of the importance of smoke and toxic fumes:

- In 1973, a passenger aboard a Varig Airlines flight (B-707) reported smoke in the lavatory shortly before the scheduled landing at Orly Airport, near Paris. Within 6 min, thick black smoke filled the cabin and cockpit. Unable to see their instruments, the pilots opened their side windows and made a forced landing in a field 4 miles from the airport. Of 135 occupants, 10 crew members and one passenger survived, all in the cockpit. The remaining 124 died from asphyxiation or the effects of toxic gases.[4,10]
- In 1980, a Saudi Airlines aircraft (L-1011) with 301 passengers and crew on board made an emergency landing after reporting an in-flight fire. The aircraft landed and taxied normally for several minutes before coming to a stop. None of the doors was opened, and all on board died.
- In 1983, a successful landing was made in Greater Cincinnati Airport after an in-flight fire aboard an Air Canada flight (DC-9). Of the 46 occupants, 23 were overcome by smoke and toxic fumes, could not leave the airplane, and died in the ensuing fire.[10]

These incidents are part of the ample evidence that many passengers in crashes or unplanned landings in which fire is involved are unable to escape from the aircraft, even though they have not sustained injuries that would prevent escape. There is a strong presumption that these passengers have succumbed to smoke or toxic fumes in the cabin. The conditions of air quality during fire emergencies are examined here, with appropriate measures that might be taken to prevent or ameliorate these conditions so as to increase the likelihood of survival and escape.

Inhibiting Ignition

Postcrash fires generally originate in one of six ways:[12]

- From release of fuel caused by wing separation during impact-survivable accidents.
- From release of fuel from damaged fuel tanks or fuel lines during impact-survivable accidents.
- From fuel tank explosions caused by external heating and other ignition sources in the crash.
- From ignition of materials in the cabin during the crash.
- In the propulsion system.
- In the landing gear system.

It is generally agreed that ignition of jet fuel constitutes the greatest potential danger in aircraft crashes.[15] In accidents in which large quantities of fuel are released and ignited (pool fires) and the integrity of the fuselage is damaged to the extent that major portions of the cabin are directly subjected to the fuel fire, the dominance of the fuel fire is clear. But even when the fuselage remains relatively intact, the radiant energy impinging on the cabin through the window ports from the flame of the pool fire is sufficient to ignite many materials.[15] Obviously, prevention of crashes and resulting fires is a major concern of the airline industry and the Federal Aviation Administration (FAA), but discussion of approaches to prevention is beyond the scope of this report.

Major fuel fires are very rare, and most incidents involving fire on aircraft involve less catastrophic situations. Although they are of considerable interest, little information is available on the progress of major past in-flight fires. However, typical origins of in-flight fires have been characterized.[12] Typical origins in the cockpit include malfunction of the electric equipment and oxygen supply system. Origins of fires in the cabin include failure in the oxygen supply system, liquid fuel spills, short circuits, matches, lighters, cigarettes and cigars, and carry on luggage. The food service galley is a common source of fires, with origins including ovens and oven exhaust systems, electric equipment, food waste storage, and the oxygen system. The lavatory is one of the few areas of the cabin where, because it is enclosed and has separate ventilation, a fire can go undetected until it reaches a dangerous magnitude. Although fires resulting from smoking in the lavatories have been of considerable concern, it appears that light wiring, speaker transformers, fluorescent-light ballasts, water heaters, and the flushing motor are more likely sources of serious lavatory fires.[12] In cargo compartments, fire sources include short circuits and cargo. Movie projectors, electric motors, and control equipment are possible causes of attic fires. Unpressurized landing gear wells—containing hydraulic fuel lines, electric controls and devices, and water-line heaters—can be sources of fire. Finally, electric equipment and avionic equipment are potential sources of electric fires.

Table 1–9 summarizes the incidents involving smoke or fumes in aircraft cabins or cockpits that have been recorded in the FAA Civil Aeromedical Institute (CAMI) Cabin Safety Data Bank. About 20 incidents are reported in a typical year, including about 13 emergency landings. An analysis of a different set of data, reports by Part 121 and Part 135 air carriers to FAA between 1980 and 1985 (summarized in Table 1–8), reveals that—of 138 incidents of fire, explosion, smoke, or related odors—68 involved mechanical failures, 25 involved electric malfunctions, 15 were galley incidents (of which eight involved spills of food or other material in the oven), 10 were lavatory fires (of which five were in waste-paper receptacles or otherwise involved paper products), eight were in the cabin (of which five involved cigarettes or lighters), and eight were categorized as "other" or "undetermined." Strict adherence to servicing codes and careful examination of such codes whenever a fire results from malfunction are required. Similarly, each crew-related fire, especially those involving cooking, should receive careful scrutiny from the point of view of equipment reliability, procedural safety, and crew performance. In large measure, the enforcement activities of FAA and the review and recommendation procedures of the National Transportation Safety Board described in Chapter 3 are intended to accomplish these aims.

Materials Testing and Selection

Failing prevention or immediate extinguishing of fire, it becomes essential to decrease the rate of flame propagation and production of toxic gas through appropriate selection of structural and decorative materials in the aircraft. The major regulatory efforts to date have been directed toward selection of minimally flammable materials for incorporation into the aircraft cabin and cockpit, in accordance with fire testing procedures noted in the Federal Aviation Regulations.[5]

On October 26, 1984, FAA published new standards that would substantially reduce the flammability of foam seat cushions;[20] transport aircraft seat cushions must meet these new standards by November 26, 1987. They require exposure of specimens of seat back and bottom cushions over a limited area to a burner with temperature and heat flux typical of cabin fire. The test specifications require that the specimens simulate the intended seat configuration and allow for the burning interaction of upholstery cover, fire blocking layer, and foam cushion material.[14] Criteria for acceptance consist of 10% allowable weight loss, burn length of 17 in., and performance essentially matching that attained by two benchmark materials.

On August 8, 1984, FAA announced proposed rules to upgrade the fire safety standards for cargo or baggage compartments in transport aircraft.[19]FAA conducted full-scale fire tests to investigate the resistance of cargo liners to flame penetration for both compartments to which crew members have access and in which fire suppression systems are required (class C compartments) and smaller compartments without access, which are designed for fire control by oxygen starvation (class D compartments). The main conclusion drawn from the testing results was that a more realistic and severe test requirement was needed for cargo liners used in both class C and class D cargo compartments.[14] The new fire test method, which measures breakthrough resistance of cargo liners, applies the maximal heat flux and temperature measured during full-scale tests under realistic ceiling and sidewall liner orientation, i.e., both vertical and horizontal. Criteria for acceptance are absence of flame penetration of ceiling and sidewall specimens and temperature above the ceiling specimen not exceeding 400°F.

In 1980, the FAA Special Aviation Fire and Explosion Reduction (SAFER) Committee recommended a specific fire scenario for FAA to use in full-scale tests and expedited the development and evaluation of the Ohio State University (OSU) rate-of-heat-release apparatus as the potential standardized test for materials.[25] On April 16, 1985, after full-scale tests, FAA announced a notice of proposed rule-making (NPRM 85–10) establishing new fire test criteria for type certification of transport aircraft that would apply to cabin interiors of all newly manufactured aircraft and all other aircraft that were type-certified after 1958.[22] In full-scale tests, various interior panel materials were subjected to situations simulating an external fuel pool fire with an open door, and the results were correlated with performance with the OSU test apparatus.[9] A panel of phenolic-fiberglass, a state-of-the-art composite used in some applications in aircraft interiors, was used as a benchmark. It added approximately 2 min to survivability, compared with other available panels studied. Criteria of 65 kW/m^2 for peak heat release rate and 65 $kW\text{-}min/m^2$ for total heat release in 2 min were established in accordance with the performance of this benchmark panel.

The Aerospace Industries Association of America (AIA), representing the airframe manufacturers, appears to have legitimate concern about the ability of the proposed NPRM 85–10 to discriminate adequately and consistently between acknowledged inferior products and molded interior components with known improved fire-resistant characteristics.[2] The proposed alternative standardized testing, advocated by AIA[1] , so discriminates, but permits use of state-of-the-art material for aircraft interior walls, ceilings, and other components that is (from a fire-resistance standpoint) only one-tenth as good as newer material, which still needs development before it can be satisfactorily used. Without some regulatory action or impetus, the use of these less developed but safer materials could be delayed until the next century. Notwithstanding a potentially adverse economic impact on airframe manufacturers, government regulators and enforcers, airline operators, and ultimately consumer-passengers, the Committee feels (with respect to all the topics under consideration here) that improved comfort and safety deserve consideration, despite the extra time that might be required in fine-tuning the product to ensure its timely incorporation into some existing and next-generation aircraft.

In general, the FAA program on flammability testing is excellent, and its research efforts to improve testing are appropriate and valuable. The Committee feels that continuing research is also needed in materials development. Although FAA standards are met by currently available materials, other materials, if developed further, would far exceed current standards and would substantially increase fire protection in aircraft. Such organizations as AIA or a similarly constituted organization of airframe manufacturers should be strongly encouraged to initiate or support programs in this field.

The Secretary of Transportation is charged under the Federal Aviation Act with responsibility for regulating air commerce in such a manner as best to promote both its development and its safety, but the Committee believes that passenger safety must be paramount. Because of the extreme hazard presented by fire and the associated smoke and noxious and toxic combustion products, minimization of fire and fumes should be of highest priority. In support of the belief that a materials development program should be encouraged, we cite the example of the National Aeronautics and Space Administration's response to the 1967 pad fire. A few simple guidelines were developed for material replacement as noted in Table 4-1, and a keyed index system was initiated. The latter consisted of an index of every nonmetallic component or individual item considered for use in spacecraft or related equipment keyed to test results for a variety of atmospheric conditions, such as odor, toxicity, total emission of organic substances, and various fire tests.

TABLE 4-1

NASA Guidelines for Selection of Replacement Materials[a]	
A.	Replace all materials that burn in 100% oxygen (3.5–16.5 psia) with nonflammable substitutes. Material substitution may not affect mission function.
B.	All products that cannot be manufactured from nonflammable materials are to be covered with an insulating, nonflammable coating to prevent the flammable substrate from being affected by heat and fire for a specified period.
C.	If A and B cannot be accomplished in a timely fashion, institute an R&D program to achieve those aims.
D.	Provide a measure of fire control by the arrangement of materials in the spacecraft. Potential flame paths can be interrupted by separating from each other items that have some propensity to burn, thus creating "fire breaks."

[a] Data from M. I. Radnofsky (personal communication).

Smoke Detection and Firefighting

On March 29, 1985, FAA added a new paragraph to the Part 121 regulations to provide that:[18]

- Each lavatory and galley in passenger-carrying airplanes be equipped with a smoke detector system or equivalent that provides a warning light in the cockpit or an audible warning in the passenger cabin that would be readily detected by a cabin attendant.
- Each lavatory be equipped with a built-in automatic fire extinguisher for each disposal receptacle for towels, paper, or waste in the lavatory.

Smoke detectors are to be installed by October 26, 1986, and automatic fire extinguishers, by April 29, 1987. The rule also increases the number of hand-operated fire extinguishers that must be carried and provides that at least two must contain Halon 1211 (bromochlorodifluoromethane) or equivalent as the extinguishing agent.

On October 10, 1985, FAA announced a proposed addition to the Part 121 regulations to require portable breathing equipment for at least one flight-attendant station in each passenger compartment and to require crew members to participate in approved firefighting drills with the portable breathing equipment.[23]

Removal of Toxic Fumes

In at least two incidents involving onboard fires, air-conditioning equipment was turned off, or engine power cut, before or after landing, and that exacerbated a serious situation with respect to toxic smoke. Smoke and toxic fumes are the principal problem in noncrash aircraft fires.

Industry practice, according to the results of the Committee's review, is to specify using maximal outside-air ventilation if smoke is present in the cabin or cockpit and turning off recirculation systems if the equipment includes this option. However, the details of the emergency procedures are inconsistent, and in some cases they cover only electric smoke or air-conditioning smoke and are not explicit regarding cabin smoke.

All procedures that the Committee reviewed specified increasing cabin altitude to 10,000 ft "to increase ventilation." Although that will increase the volume of air flowing through the cabin, the lower pressure will also increase the volume of smoke produced by a given fire, and there would be little or no reduction in smoke concentration. Any reduction in burning rate due to the decrease in partial pressure of oxygen in the cabin is insignificant.

Deployment of oxygen masks in a fire is not recommended by the airline procedures, because current passenger oxygen masks only increase the oxygen concentrations in the air provided to passengers and do not reduce the hazard of toxic fumes. The use of oxygen masks could thus lead to an unwarranted sense of security.

Details of the use of ventilation and pressurization during an emergency descent due to a cabin fire were not found in any of the procedures reviewed. In the Air Canada incident near Cincinnati, in which air-conditioning was turned off, air entering the cabin during descent probably forced smoke into the cockpit. During an unpressurized rapid descent, air enters the cabin through the negative relief valves, which are installed to prevent crushing of the fuselage. On the DC-9, which was involved in the Air Canada incident, the negative relief valves are above the ceiling in the aft pressure bulkhead. Air entering through these valves would have been forced over the fire and would have carried smoke forward through the area above the ceiling and caused it to enter the forward cabin and the cockpit.

As discussed in Appendix A, the steady-state concentration of fumes in an aircraft cabin is the effective volume production rate, P, divided by the loss frequency, L. L is found by dividing the outside-air ventilation rate by the cabin volume. Both decreasing P (by inhibiting ignition, extinguishing a fire, and improving materials) and increasing L (by using the maximal available outside-air ventilation rate) reduce fume concentration. The Committee strongly recommends that cabin-fire instructional material emphasize the need to turn on all available air packs to full volume if a cabin fire or smoke is present. That should reduce the possibility of smoke in the cabin and increase the likelihood of passenger survival. All recirculation should be turned off if the equipment includes this option.

FAA, manufacturers, and the airlines should conduct further analyses aimed at developing detailed procedures for optimal crew management of pressurization, ventilation, auxiliary ventilation (if available), and exhaust systems to control smoke during an emergency descent. Because current supplemental oxygen masks are designed to substitute cabin air for oxygen automatically whenever the cabin pressure is below the equivalent altitude of about 18,000 ft, even attaching the oxygen masks to a clean-air intake would not eliminate the hazard of breathing smoke and toxic fumes. Detailed procedures should also be developed for engine and air pack shutdown and for use of an auxiliary power unit after landing, to continue to provide ventilation and prevent buildup of heat and flammable fumes that could produce flashover during evacuation.

Individual Smoke and Fume Protection

Because smoke and toxic fumes are principal causes of death in survivable crashes, smoke hoods and other passenger breathing devices have been proposed as a way of protecting passengers and increasing the likelihood of their survival. The studies referred to below suggest that some protection could be gained through their use, but there are limitations and difficulties.

After a crash in 1965, CAMI embarked on a program to develop passenger smoke hoods.[17] The program led to announcement of an amendment to FAR Part 121 in 1969 that would have required protective smoke hoods to be available on all civil air carrier.[24] A number of critical comments were received, mostly involving hood safety, practicality, slowing of evacuation, and justification of the specifications. In response to these comments, and over the strong objection of the medical and regulatory arms of FAA, the proposed rule was withdrawn in September 1969.[17]

Several protective devices have been developed, ranging from a simple moist multilayer cloth large enough to cover the mouth and nose and held to the mouth and nose by hand or by an elastic band around the head (the North American Rockwell smoke mask) to hoods incorporating compressed-air or oxygen generators (e.g., the experimental FAA-Sheldahl hood with self-contained air supply, the Lear-Siegler air capsule, and the Scott aviation emergency smoke hood and breathing device). In addition, devices developed for other purposes, such as escape from mines, have been examined for their applicability as passenger protective devices.[17]

It was widely publicized that cabin attendants passed out wet towels during the Air Canada in-flight fire aboard a DC-9. However, that was probably not effective—only a small percentage of the passengers who were given wet towels survived.[3]

The relative advantages and disadvantages of passenger smoke hoods and other protective breathing devices have been assessed. In 1969, the Air Transport Association appended to its comments on the proposed rule the concerns of Richard L. Riley and Solbert Permutt, of the Johns Hopkins University Department of Environmental Medicine, about the hazard of hypoxia created by the configuration of the smoke hood itself. They were especially concerned about prolonged breath-holding and were uncertain about whether all passengers would remove their hoods when the carbon dioxide in the hoods exceeded the generally accepted safe concentration.

In 1970, FAA asked the National Research Council Space Science Board to evaluate the smoke hood.[13] The Board pointed out several potential hazards, including the narcotic effect of high concentrations of carbon dioxide (9.2%), the impossibility of effecting resuscitation once respiratory failure has been brought about by inhalation of pure carbon dioxide, and the possibility of hypoxia. It raised the legal question regarding a lethally injured person who is found wearing a smoke hood after a fire when cause of death is difficult to determine. And it raised questions about the use of the hood by people with cardiac disease or pulmonary dysfunction and about the fitting of the device for infants, children, and people with an abnormal neck size.

In 1976, several smoke hoods were reviewed in a report of the NATO Advisory Group for Aerospace Research and Development.[17] The report examined leakage, effectiveness in toxic environments, vision, acoustic attenuation, effectiveness in dense smoke environments, and effectiveness of safety briefings. It concluded that the available Sheldahl rebreathing smoke hood with septal neck seal (Type S) "can provide protection from smoke, toxic fumes, and flame in postcrash fire emergency egress" and stated that "its demonstrated merits far outweigh any potential risks or problems." That judgment appears to have considered all the problems noted above, except the issues of legal responsibility and liability.

The Ontario Research Foundation, in Canada, recently completed an evaluation of over 20 devices and extensive testing of six, all of which include filtration or absorption.[6] Test criteria included edge leakage, smoke and toxic gas penetration of filtration units, condition of inhaled air (carbon dioxide, oxygen, and temperature), comfort, ease of donning, breathing resistance of filtration units, vision and communication, resistance to flame, cost, size and weight, and compatibility with current passenger supplemental oxygen systems. The report concluded that a compact device providing both depressurization and protection from toxic smoke and gas for airline passengers is feasible and that the devices tested can provide several extra minutes of escape time. However, although the study considered use of the devices under several different conditions (sitting, walking, talking, and light exercise), it did not evaluate their use under conditions corresponding to the evacuation of an aircraft during a fire.

The FAA position is that efforts to reduce the likelihood of ignition or smoldering fire have diminished the need for individual passenger smoke and fume protection devices. FAA bases its position on the relative merits of four basic types of passenger emergency breathing devices: simple smoke hoods with neck seal and no oxygen supply, hoods or masks that connect to the individual ventilation outlets (gaspers), modifications of current oxygen masks, and hoods or masks with individual self-contained oxygen supplies.

- FAA concluded that simple smoke hoods are of limited utility, because of the restrictions on the length of time they can be worn before effects of hypoxia, carbon dioxide poisoning, etc., set in. The time involved in a typical incident associated with an in-flight fire at cruise altitude—including emergency descent, landing at the (possibly unscheduled) airport, stopping, and evacuation—is sufficient to make adverse behavioral and physiologic effects likely. In one study in which smoke hoods were donned in a darkened cabin and the aircraft was evacuated, the use of smoke hoods reportedly increased evacuation times by 50% (T. E. McSweeny, personal communication, 1986).
- Not all aircraft have gaspers, and they are not commonly selected by airlines for current aircraft

models. Thus, the connection of hoods or masks to gaspers cannot be considered a general solution. More important, most ventilation systems on current aircraft involve at least some recirculation of air in the cabin. Connection of hoods or masks to gaspers thus presents the possibility of introducing smoke or toxic fumes directly into the passengers' air supply.

- FAA considers modification of current supplemental oxygen masks to be the most promising of the options examined. However, the current diluter demand mask operates as a function of cabin altitude. Below a cabin altitude of about 20,000 ft, no oxygen is introduced, and the system relies on air from the standard ventilation system, so it is also subject to possible contamination with smoke and fumes, as are gaspers. Modification of the oxygen supply system to cover the time required for descent, landing, stopping, and evacuation would require re-engineering of the oxygen supply systems and considerable extension of the oxygen supply. Careful thought must be given to the addition of large amounts of oxygen in an extensive network of overhead tubes, because it would be in the same portion of the cabin that is typically subjected to the greatest temperatures and to flashover conditions.
- FAA has given less attention to self-contained breathing devices, mostly because of their greater cost. Furthermore, passengers have found it difficult to don and use current oxygen masks and life vests properly and would probably have even more trouble with more complicated breathing devices.

For these reasons, FAA has chosen to pursue engineering solutions involving selection of materials, fire detection and extinguishing, evacuation and development of a method of purging the aircraft of smoke and toxic fumes in flight, rather than passenger protective breathing devices.[11]

The FAA policy, however, is based on the premises that flashover is not survivable and that the primary concern in postcrash fires or in-flight fires once the aircraft has landed and stopped is rapid evacuation. Although this initially appears valid, some people have survived flashover. For example, two passengers and an attendant hid in the aft stairwell of a B-737 involved in a large postcrash fire in 1965 until rescued 25 min later.[16] They survived, one on top of the other, by breathing through a small crack in the fuselage. The one on top, a 61-yr-old man, died of burn injuries on the seventh postcrash day. In the recent British Airtours accident in Manchester, England, a B-737 caught fire and burned. A 14-yr-old boy was removed from the aircraft approximately 5 min after flashover and survived. An older man was rescued from the same aircraft 30 min after flashover and survived for 6 d before succumbing (E. J. Trimbell, personal communication, 1986). On the basis of that incident, the Accident Investigation Branch of the U.K. Department of Transport has strongly recommended that the Civil Aviation Authority require passenger breathing devices in that country.

The Committee feels that passenger smoke hoods and breathing devices should be evaluated in terms of their potential contribution to survival and their effect on such factors as evacuation time. In case toxic fumes are the reason for a need for quick escape, protection at the slight expense of speed might save many lives. This needs to be critically examined.

Despite the incompleteness of data on the effectiveness of passenger protective breathing devices under realistic conditions, the Committee recommends that such systems be studied. Published reports suggest that one passenger could be saved for each second added to the time available for escape in an emergency evacuation of an aircraft on which a fire is generating toxic smoke.[7] It might be worth while to reinvestigate this life-sustaining protective breathing equipment in light of recent developments in contaminant absorption and self-contained sources of air or oxygen for such units.

Emergency Escape

On October 26, 1984, FAA published a new requirement for floor-proximity emergency escape-path marking that provides visual guidance for emergency escape when all sources of cabin lighting more than 4 ft above the floor are totally obscured.[21] Although this standard does not affect cabin air quality, it is designed to deal with a situation of severely degenerated air quality.

Passenger Safety Briefings

In-flight or postcrash fires are never mentioned by cabin crew in their passenger safety instructions. Toxic, noxious, and blinding gaseous products and particulate matter resulting from fire stratify in the aircraft in such a way that the best air is closest to the floor, but this potentially vital information is not given to passengers. However, when speed is critical for evacuation, staying close to the floor under conditions of limited visibility might be counterproductive.

DEPRESSURIZATION

The primary threat to the passenger in depressurization is hypoxia. The main problem is in inducing passengers to don their oxygen masks correctly and quickly.

Records in the CAMI Cabin Safety Data Bank show a total of 355 incidents involving depressurization in 1974–1983 (Table 4-2). Of these, 43% were classified as "significant" incidents—i.e., cabin pressure decreased to an equivalent altitude above 14,000 ft, passenger masks were deployed, or an injury resulted.[8] Only one death occurred: that of a passenger with a history of heart problems. There were three serious injuries: one cockpit crew member had a broken arm, one passenger had a nonfatal heart attack, and one passenger had a collapsed lung. Sixty-six passengers and two cockpit crew members reported minor ear pain; 55 passengers and two cockpit crew members reported intense ear pain; 11 passengers incurred serious ear damage, including eight with bleeding ears; and three passengers suffered nosebleeds. No flight attendants reported any of these problems.

Seventeen cases of hypoxia were reported: seven passengers and five flight attendants suffered mild hypoxia, and one passenger and four flight attendants suffered loss of consciousness. No cockpit crew members reported symptoms of hypoxia, perhaps because they are usually the first to be aware that depressurization has occurred and have ready access to masks with demand regulators.

TABLE 4-2 Ten-Year History of Reported Depressurizations, 1974–1983a

Year	No. Significant[b] Incidents	No. Minor Incidents	No. Undefined Incidents	Total No. Incidents
1974	18	24	12	54
1975	15	22	1	38
1976	17	10	6	33
1977	16	13	2	31
1978	16	17	6	39
1979	20	18	7	45
1980	19	20	5	44
1981	18	12	4	34
1982	8	8	2	18
1983	7	6	6	19
Total	154	150	51	355

[a] Data from Higgins.[8]

[b] See text for definition of "significant."

In studies conducted at CAMI in 1976, it was determined that the physical activity typical of a flight attendants duties reduces the time of useful consciousness (amount of time until mental functioning deteriorates) by about 40% compared with that of an inactive passenger. If rapid depressurization occurs, an attendant has only 15–20 s, depending on altitude and final cabin pressure, to don a mask before adverse effects, such as mental sluggishness, begin.

The continuous-flow passenger mask has proved satisfactory for cabin altitudes up to 40,000 ft, when properly used. Most of the problems with this type of mask appear to be associated with the lack of timely or proper donning—for example, failure to pull the mask down to activate the oxygen flow, failure to ensure that the mask covers both nose and mouth, and failure to tighten the straps to ensure a good fit.[8]

The problem of depressurization thus has to do essentially with passenger retention of safety instructions and quickness of response. Issues involving cabin safety procedures and information presentation are dealt with in Chapter 3.

CONCLUSIONS AND RECOMMENDATIONS

The Committee concludes that, although the ignition and propagation of fires and the resulting generation of combustion products aboard commercial airliners is complex, much can be accomplished toward alleviating the associated hazards. The use of materials that have high resistance to burning, that will not propagate a flame, and that will not generate toxic products when subjected to heat loads sufficient to cause currently used materials to degenerate would constitute a distinct improvement in passenger safety and air quality in the event of an in-flight, postcrash, or landing fire. The Committee recommends that FAA review current airline operating procedures and flight crew instructions for emergencies involving cabin fire or smoke; this review should cover every type of aircraft, regardless of size, in commercial service in the United States. The Committee recommends that the Aerospace Industries Association of America, a similarly constituted organization of airframe manufacturers, or even an individual manufacturer be encouraged to fund and initiate a program to develop a more fire-resistant set of materials from which to fabricate fully functional interior materials for aircraft.

The Committee concludes that smoke hoods or other protective breathing and vision devices would provide additional passenger survival time in an otherwise debilitating situation that might normally preclude survival. FAA should re-evaluate smoke hoods or special breathing devices for passenger use.

The Committee recommends immediate implementation of directions to turn on all air packs and to turn off internal recirculation systems in case of onboard fire. The Committee recommends that air contamination modeling studies and confirmatory live testing in aircraft be performed as soon as possible, by properly constituted FAA-industry teams to determine conclusively the advantages of turning on all packs to full volume when there is an in-flight fire.

The Committee recommends that FAA require information on proper response to fire emergencies to be included in oral and written passenger safety briefings.

REFERENCES

1. Aerospace Industries Association of America, Inc. Comments to NPRM 85-10a, "Improved Flammability Standards for Materials Used in the Interiors of Transport Category Airplane Cabins," Docket No. 24594. (unpublished communication, October 16, 1985)
2. Aerospace Industries Association of America, Inc. Final AIA comments to NPRM 85-10A, "Improved Flammability Standards for Materials Used in the Interiors of Transport Category Airplane Cabins," Docket No. 24594. (unpublished communication, January 8, 1986)
3. Barthelmess, S. Flight 797: A human-factors perspective. Part II. FSF Flight Safe. Dig. (Aug.):12–17, 1984.
4. Bulloch, C. Survivability in aircraft fires: New standards are needed. Interavia 34:557–558, 1979.
5. Compartment interiors. Code of Federal Regulations, Title 14, Pt. 25.853. Washington, D.C.: U.S. Government Printing Office, 1985.
6. Dranitsaris, P. An Evaluation of Candidate Smoke Masks for Passenger Use in Airplane Fires. Draft Report P-4612/G. Mississauga, Ontario: Ontario Research Foundation, 1986.
7. Hall, J. R., and S. W. Stiefel. Decision Analysis Model for Passenger Aircraft Fire Safety with Application to Fire-Blocking of Seats: Interim Report, November 1982–December 1983. NBSIR 84-2817, DOT/FAA/CT-84/8. Atlantic City, N.J.: U.S. Federal Aviation Administration Technical Center, 1984.

8. Higgins, E. A. Protective breathing: Oxygen mask use/problems, pp. 61–65. In Flight Safety Foundation, Inc. Proceedings of Cabin Safety Conference and Workshop, December 11–14, 1984. Washington, D.C. : U.S. Federal Aviation Administration, Office of Aviation Safety, 1985.
9. Hill, R. G., T. I. Eklund, and C. P. Sarkos. Aircraft Interior Panel Test Criteria Derived from Full-Scale Fire Tests. DOB/FAA/CT-85/23. Atlantic City, N.J.: U.S. Federal Aviation Administration, Technical Center, 1985.
10. Johnston, W. L., and P. T. Cahalane. Examination of fire safety of commercial aircraft cabins. SAFE J. 15(2):4–9, 1985.
11. McSweeny, T. E. FAA cabin safety activities, pp. 30–44. In Flight Safety Foundation, Inc. Proceedings of Cabin Safety Conference and Workshop, December 11–14, 1984. Washington, D.C.: U.S. Federal Aviation Administration, Office of Aviation Safety, 1985.
12. Miniszewski, K. R., T. E. Waterman, J. A. Campbell, and F. Salzberg. Fire Management/ Suppression Systems/Concepts Relating to Aircraft Cabin Fire Safety: Final Report. DOT/ FAA/CT-82/134. Atlantic City, N.J.: U.S. Federal Aviation Administration Technical Center, 1983.
13. National Research Council, Space Science Board. Evaluation of Smoke Hoods. 1970. (unpublished document)
14. Sarkos, C. P. New FAA regulations improve aircraft fire safety. ASTM Standardization News 13 (12):33–39, 1985.
15. Sarkos, C. P., R. G. Hill, and W. D. Howell. Full-scale wide-body test article employed to study post crash fuel fires. ICAO Bull. 37(10):30–39, 1982.

16. Snow, C. C., J. J. Carroll, and M. A. Allgood. Survival in Emergency Escape from Passenger Aircraft. AM 70–16. Washington, D.C.: U.S. Federal Aviation Administration, Office of Aviation Medicine, 1970. (available from NTIS as AD-735388)
17. Snyder, R. G. Advanced Techniques in Crash Impact Protection and Emergency Egress from Air Transport Aircraft. AGARD-AG-221. London: North Atlantic Treaty Organization, Advisory Group for Aerospace Research and Development, 1976. (available from NTIS as AD-A029375)
18. U.S. Federal Aviation Administration. Airplane cabin fire protection: Final rule. Federal Register 50(29 March):12726–12733, 1985.
19. U.S. Federal Aviation Administration. Fire protection requirements for cargo or baggage compartments: Notice of proposed rulemaking. Federal Register 49(8 Aug.):31830–31838, 1984.
20. U.S. Federal Aviation Administration. Flammability requirements for aircraft seat cushions: Final rule. Federal Register 49(26 Oct.):43188–43200, 1984.
21. U.S. Federal Aviation Administration. Floor proximity emergency escape path marking: Final rule. Federal Register 49(26 Oct.):43182–43186, 1984.
22. U.S. Federal Aviation Administration. Improved flammability standards for materials used in the interior of transport category airplane cabins: Notice of proposed rulemaking. NPRM 85-10. Federal Register 50(16 April):15038–15053, 1985.
23. U.S. Federal Aviation Administration. Protective breathing equipment: Notice of proposed rule-making. Federal Register 50(10 Oct.):41452–41459, 1985.
24. U.S. Federal Aviation Administration. Protective smoke hoods for emergency use by passengers and crewmembers: Notice of proposed rule making. Federal Register 34(11 Jan.):465–466, 1969.

25. U.S. Federal Aviation Administration, Special Aviation Fire and Explosion Reduction (SAFER) Advisory Committee. Final Report. Vol. 1. FAA-ASF-80-4. Washington, D.C.: U.S. Federal Aviation Administration, 1980. (available from NTIS as AD-A092016)

5

Cabin Air Pollutants: Sources and Exposures

Little is known about the environment in the passenger cabins of commercial aircraft under routine flight conditions, and what is known is limited in scope. Relationships among source strengths of pollutants, physical factors (such as ventilation rates and operating modes), occupancy loads, and activities (such as eating and smoking) have not been systematically studied. Lacking a repository of the existing information, the Committee searched the published literature to obtain relevant material on pollutants known to be potentially hazardous or to cause acute irritation and on physical factors that affect comfort. On the basis of the results of the searches, this chapter discusses ozone, cosmic radiation, ground fumes, tobacco smoke and carbon monoxide, biologic aerosols, relative humidity, cabin pressure, carbon dioxide, volatile organic chemicals, and pesticides.

OZONE

Ozone in Commercial Aircraft Cabins

Ozone is present in the atmosphere as a consequence of the photochemical conversion of oxygen by solar ultraviolet radiation. A marked and progressive increase in ozone concentration occurs between the tropopause and the stratosphere—i.e., it occurs within the flight altitude of commercial aircraft.

The mean ambient ozone concentration increases with increasing latitude, is maximal during spring, and often varies with weather systems to result in high ozone plumes descending down to lower altitudes.

In the early 1960s, R. I. Brabets et al.[24] established that jet aircraft operating in the stratosphere encountered ozone and that it was only partially removed from the internal environment of the aircraft by the compression-ventilation system. In response to these findings, the Global Atmospheric Sampling Program (GASP), started by the National Aeronautics and Space Administration in 1977, measured ozone concentrations in the cabins of two commercially operated aircraft. In 1980, Nastrom et al.[107] reported that over 5,600 observations were made in this project in a B-747-100 and a B-747-SP. The ozone concentrations measured in the outside air and in the cabin of an unmodified B-747-SP are shown in Figure 5-1.

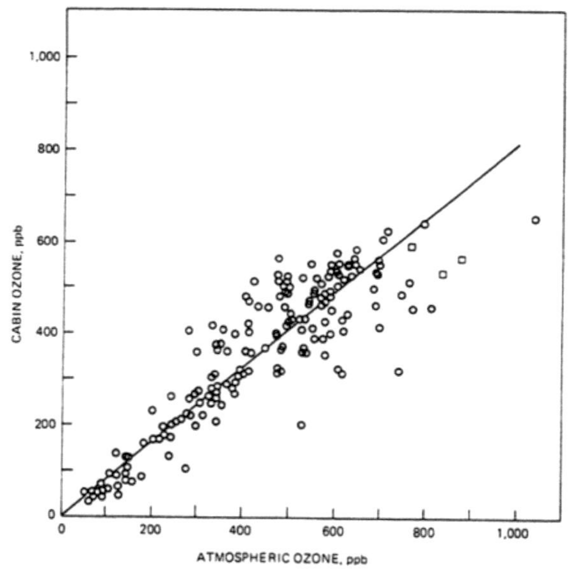

FIGURE 5-1
Correlation (slope, 0.82) of cabin with atmospheric ozone mixing ratios. Data were obtained during April, May, and June 1977 before changes were made in B-747-SP air circulation system. Squares show data taken in April. Reprinted from Perkins et al.[114]

Crew members' and passengers' complaints of physical discomfort on high-altitude flights led the Federal Aviation Administration (FAA) to begin to collect information on possible causes.[159] In 1977, the agency took five steps to investigate further whether ozone was the pollutant responsible for the complaints:

- It published an advisory circular that defined ozone irritation, discussed its cause and symptoms, and described means of dealing with it.[159]
- It initiated a research project in the Civil Aeromedical Institute to study the health effects of exposure to ozone in the aviation environment.
- It issued Advance Notice of Proposed Rulemaking No. 7$1$2$32 to seek information concerning ozone.[157]
- It initiated a project to measure the constituents of the upper atmosphere.
- It initiated a study of available data on ozone concentrations at flight altitudes to provide an estimate of average atmospheric ozone at flight altitudes.

On the basis of these efforts, FAA established a standard for cabin ozone concentration.[31] The Code of Federal Regulations of January 1, 1985, stated the following: "The airplane cabin ozone concentration during flight must be shown not to exceed 0.25 ppm, sea level equivalent, at any time above flight level 320 [32,000 ft at standard atmosphere]; or 0.10 ppmv during any 3-hour interval above flight level 270 [27,000 ft at standard atmosphere]."

Health Effects of Ozone under High-Altitude Conditions

The following text discusses several experimental studies involving humans. See Chapter 6 for discussion of findings on human exposure and resulting effects during flight.

Toxic effects of ozone on the respiratory system have been investigated in numerous human studies involving controlled exposures to ozone at concentrations observed in community air.[156] The characteristic odor of ozone can be detected by some people exposed to it at concentrations as low as 0.001 ppm.[143] This may be important because of perception of exposure. The threshold varies among individuals, but most people can detect ozone at 0.02 ppm. Controlled human studies have reported respiratory symptoms and significant decrements in pulmonary function associated with ozone exposure. The severity of reported symptoms generally parallels the observed impairment in pulmonary function. Symptoms include cough, upper airway irritation, tickle in the throat, chest discomfort, substantial pain or soreness, difficulty or pain in taking a deep breath, shortness of breath, wheezing, headache, fatigue, nasal congestion, and eye irritation. Cough is the symptom most strongly correlated with the decrement in pulmonary function. These symptoms and the alteration in pulmonary function usually disappear soon after the termination of the exposure. Some subjects have reported persistence of changes in excess of 24 h, but most disappear within 2–4 h. If exposure is repeated within 24–48 h, pulmonary function decrements are markedly greater.[65]

Studies in environmental chambers using at-rest (i.e., no-exercise) exposures to ozone have shown that ambient ozone at 0.5 ppm or more induces significant decrements in pulmonary function.[66] Impairment in pulmonary function occurs at much lower ambient concentrations of ozone if subjects are exercising. Subjects engaged in light exercise (ventilation, approximately 20–25 L/min) had significant pulmonary function decrements when ozone was present at 0.37 ppm. In persons exercising moderately to heavily (26–40 L/min), pulmonary decrements have been observed during exposures at 0.14–0.18 ppm.[50 84 97]

Lategola and associates attempted more quantitative evaluation of problems associated with ozone exposures of flight attendants and passengers. Lategola et al.[85] exposed 55 young subjects (29 men and 26 women) to ambient air and to an ozone environment in an altitude chamber maintained at 1,829 m (6,000 ft). Subjects served as their own controls in each experiment. Two major experiments were conducted on 27 subjects (15 men and 12 women) and 28 subjects (14 men and 14 women).

In the first experiment, ozone concentrations* were 0 and 315 $\mu g/m^3$(0.0 and 0.2 ppm), exposure time was 4 h (with four 10-min exercise periods, the first three at lower levels of activity and the fourth at a higher level), and pulmonary function and subjective evaluations were noted before and after exposure. Pulmonary function and subjective responses were recorded near sea level before and 10 min after the altitude exposures. Other studies—on vision, hand steadiness, and memory—were conducted during the high-altitude exposures. Men exercised at ventilation of 20 L/min in the first three exercise periods and 30 L/min in the last period, just before descent; women exercised at 13 and 17 L/min, respectively. No alterations in measured pulmonary functions were found; although slight discomfort was reported, it was not significantly related to ozone exposure. In the second experiment, the ozone concentration was 475 $\mu g/m^3$ (0.3 ppm), and only three exercise periods were used. Men exercised at 24.9 L/min in the first two periods and 38.6 L/min in the last, and women at 16.4 and 20.9 L/min, respectively. Significantly greater symptom scores were found after the last exercise period and after termination of the experiment. In this experiment, differences between the no-ozone and ozone responses in all spirometry measures—forced vital capacity (FVC), forced expiratory volume (FEV_1), and forced expiratory flow ($FEF_{25-75\%}$ and $FEF_{75-95\%}$)—in each sex group were statistically significant ($p < 0.05$). The two lung-volume measures manifested smaller changes than did flow-rate measures. Symptom scores were greater in men than in women during the last exercise (treadmill) period, but the difference was not statistically significant. The results indicate increased symptoms and pulmonary function decrements among normal subjects at 0.3 ppm, but not at 0.2 ppm with light exercise.

* Note that, as ambient pressure decreases at high altitude, ozone concentration remains the same when expressed in parts per million, but decreases in proportion to increasing altitude when expressed in micrograms per cubic meter. Therefore, knowledge of atmospheric pressure and temperature is generally needed for correct conversion of ppm readings to $\mu g/m^3$ concentrations.

Lategola et al.[86] also studied 40 middle-aged men—20 smokers and 20 nonsmokers—exposed in an altitude chamber (1,829 m) while resting for 3 h in environments containing ozone at 0 or 475 $\mu g/m^3$ (0.0 or 0.3 ppm). Eye discomfort was the most frequently reported symptom; headache and nose and throat irritation were also reported. All subjects combined manifested small but statistically significant decrements in FVC, FEV_1, and $FEF_{75-95\%}$, primarily owing to changes in the nonsmoking group. Smokers reported fewer or less severe symptoms, in confirmation of observations reported by others. The study tended to confirm small but significant respiratory effects at 0.3 ppm among nonsmoking normal adults under high-altitude conditions. The ozone concentrations used in the Lategola et al. studies were, however, generally lower than those reported to occur in some aircraft at high altitudes.

Determination of the effects of known aircraft cabin ozone concentrations on passengers and flight attendants will require additional information from studies conducted on board, as well as immediately after flights, with continuous measurements of the cabin environment.

Groups at Increased Risk of Health Effects

Epidemiologic investigations of high-risk groups have played a predominant role in the development of the current ambient air quality standard for ozone. As far back as 1961, Schoettlin and Landau[135] studied 137 asthmatics in the Los Angeles basin during a 3-mo period when high oxidant concentrations due to smog were anticipated. They found a statistically significant increase in the number of mild attacks when peak oxidant concentrations exceeded 0.25 ppm. A further assessment by Heuss et al.[59] associated these asthmatic attacks with hourly average concentrations as low as 0.15 ppm. They concluded that, when the ozone concentration is 0.15 ppm, there is a 1% chance of a 5% increase in asthmatic attacks. Barth et al.[21] extrapolated these data and concluded that there "is a likelihood of an increased asthmatic attack incidence for very sensitive patients at levels well below 0.15 ppm rather than just a chance of a small increase in attacks at the 0.15 ppm level."

Recommendations

Chapter 3 pointed out that the federal regulations concerning aircraft cabin ozone concentrations may be complied with either through the use of air treatment equipment (usually a catalytic converter) or through the choice of routes and altitudes that avoid areas of high ozone concentration. Ozone concentrations in aircraft depend also on latitude, not only on altitude. In 1978–1979, FAA monitored ozone on flights (mostly at 30,000–40,000 ft) and found that 11% were in violation of FAA's ozone concentration limits.[123] Because catalytic converters are subject to contamination and loss of efficiency, it is suggested that FAA establish policies for periodic removal and testing, so that the effective life of these units can be established. A program of monitoring is needed, to establish compliance with the existing standard and to determine whether the catalytic converters are operating normally and effectively. These data should be maintained in such a manner that they can be used for reference on passenger and crew exposures to ozone and to document the concentrations of ozone.

COSMIC RADIATION

We are exposed to ionizing radiation from several sources. Some is natural, such as cosmic radiation and terrestrial radiation, and some is from man-made sources, such as medical x rays, radioisotope drugs, nuclear fallout, nuclear power-plant emission, uranium and phosphate mine tailings, and nuclear waste materials. The question before the Committee is whether the incremental exposure of passengers and crew of commercial subsonic aircraft results in an unacceptable risk.

Characteristics of Cosmic Radiation

Cosmic radiation is both solar and galactic in origin. Galactic radiation is composed of protons (87%), alpha particles (11%), a few nuclei with atomic number of 3 or more (approximately 1%), and electrons at energies up to 10^{20} eV (approximately 1%). The normative range of energies is 10^{8}–10^{11} eV. The sun generates a continuous flux of lower-energy (approximately 10^{3} eV charged particles, and occasional solar magnetic disturbances generate large quantities of particles with energies up to several billion electron volts; the typical range is 1–100 MeV. The integrated flux of solar particles with energies of 20 MeV or more to the top of the earth's atmosphere varies with the 11-yr solar cycle between 10^5 and 10^{10} particles/cm^2 per year. The integrated flux of galactic particles is more constant, at about 10^8 particles/cm^2 per year.

These primary solar and galactic particles are almost completely attenuated as they penetrate the atmosphere down to an altitude of about 20 km (65,600 ft). However, as they pass through an increasingly dense atmosphere, they undergo nuclear interactions. Hence, at the altitude of 20 km only 50% of the original protons, 25% of the original alpha particles, and 3% of the heavier nuclei are left. But there is a buildup of secondary particles—neutrons, protons, and pions. Further pion decay produces electrons, photons, and muons. As a result, there is a cosmic radiation maximum at 20 km. A net attenuation in particle flux density occurs at lower altitudes, reducing both the number and the energy of secondary particles produced. At altitudes below 6 km (19,700 ft), muons and associated decay electrons are the dominant components of the cosmic-ray particle flux. Figure 5-2 illustrates the components of cosmic radiation dose equivalent rates as a function of altitude.

Secondary particles react with tissue through several mechanisms, including ionization (stripping of electrons) and direct inelastic and elastic collisions with nuclei. Both protons and gamma rays can interact with electrons and cause ionization of molecular structures in tissue. The heavier neutrons can have elastic collisions with lighter elements in tissue. Because of the abundance of the hydrogen nucleus in tissue, it is the most likely target nucleus for elastic scattering. Some of the energy is lost as gamma photons in inelastic collisions with heavier target nuclei. In both types of collisions, the now-energized target nucleus penetrates tissue as an ionizing particle. Like directly ionizing proton particles, these recoil protons are massive, compared with electrons, and dissipate energy over a relatively short path. Thus, the biologic effectiveness of radiation depends on the characteristics of the radiation, and not only on its energy. Because muons and associated fast electrons are essentially unattenuated by the body, the dose equivalent rate, in millirems per hour, as a function of altitude is determined essentially by the flux of protons and fast neutrons. The flux rates for fast neutrons at various altitudes are shown in Figure 5-3.

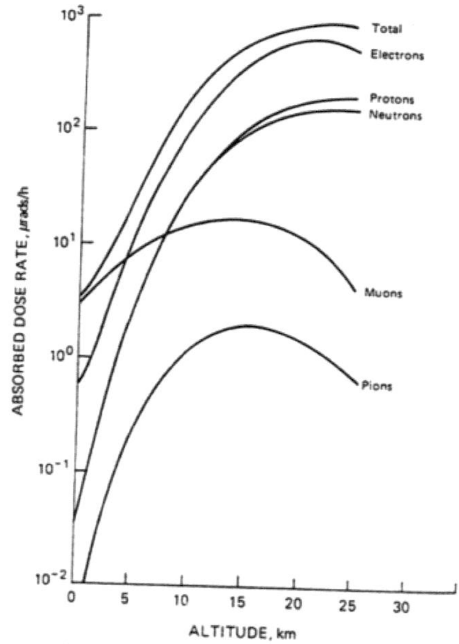

FIGURE 5-2
Absorbed dose rates at depth of 5 cm in 30-cm-thick slab of tissue from various components of cosmic radiation at solar minimum and at geomagnetic latitude of 55 degrees N. Reprinted with permission from National Council on Radiation Protection and Measurements.[109]

The dose equivalent rate of cosmic radiation in millirems per hour as a function of altitude is illustrated in Figure 5-4. The equivalent dose varies temporally (with time of maximal solar activity) and with latitude. The spatial and temporal variations have been determined from several direct-measurement programs conducted during the late 1960s and early 1970s. At altitudes typical of subsonic commercial aircraft, 9–12 km (29,500–39,400 ft), the cosmic-ray dose equivalent rate is approximately 100 times the rate at sea level. The newer, higher-performance aircraft are certified to 46,000 ft (14 km). The cosmic-ray dose equivalent rate at 14 km is nearly twice the rate at 10 km (32,800 ft).

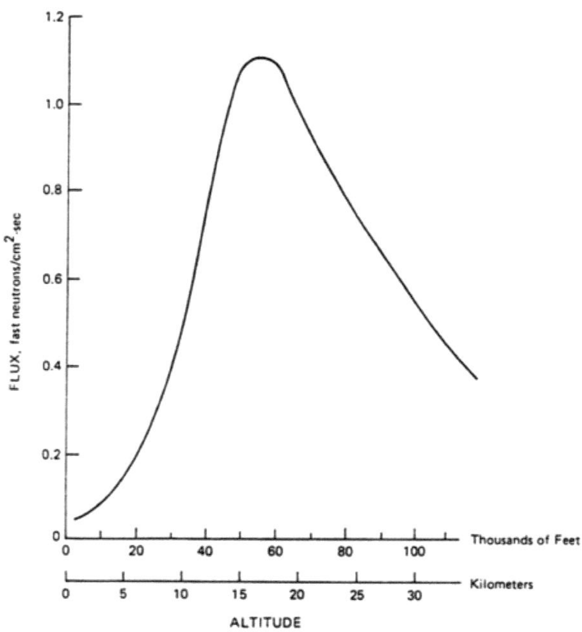

FIGURE 5-3
Altitude profile of atmospheric neutron flux. Adapted from Schaefer.[131]

Variation in solar activity and the interaction of charged particles in the earth's magnetic field result in higher cosmic radiation flux at higher latitudes and during solar flares. Figure 5-5 illustrates the profiles of dose equivalent rates by altitude, latitude, and solar-flare activity.

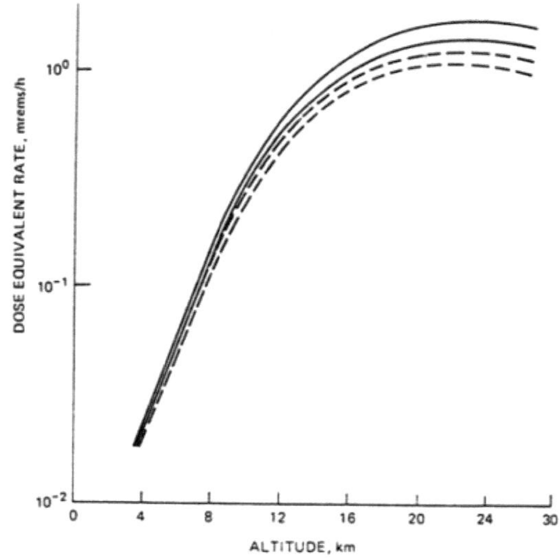

FIGURE 5-4
Total cosmic-ray dose equivalent rate at 5-cm depth in 30-cm slab of tissue at $gamma_m$ = 55 degrees N (-----) and 43 degrees N (-----) at solar minimum (upper curve) and solar maximum (lower curve). Quality factors for neutrons as function of energy are included in calculations. Reprinted with permission from National Council on Radiation Protection and Measurements.[109]

Exposure of Passengers and Crew

From Figure 5-5, it is relatively easy to estimate the dose equivalent exposure for a particular flight or for an individual. A 5-h trans-Atlantic flight at midlatitude and an altitude of 12 km (39,400 ft) might result in an equivalent whole-body dose of 2.5 mrems. If the same flight goes over the pole during a time of more intense solar activity, the dose equivalent might be 10 mrems. In general, the hourly does rate at a jet cruising altitude is approximately 100 times the ground-level rate. A person who lived near sea level would have to spend about 200–600 h/yr at cruising altitude to double his or her exposure to cosmic radiation.

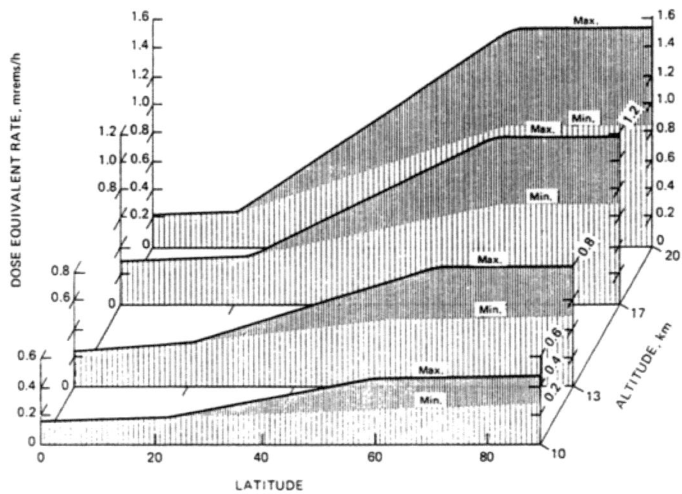

FIGURE 5-5
Best values for maximal and minimal galactic dose equivalent rate as function of latitude and altitude. Reprinted with permission from Baily.[18]

Wallace and Sondhaus[165] calculated the cosmic radiation exposure to passengers and crew in subsonic commercial travel. The database was for the year 1974 and was limited to domestic and overseas flights longer than 322 km (200 miles). The calculations were made for U.S. residents on the basis of some simplifying assumptions. A complex model was developed according to aircraft type, flight crew and passenger capacities, climb rates, cruise speeds, and flight paths. Matrices were developed for neutron and secondary charged-particle densities according to latitude, altitude, and solar conditions. The Aircraft Radiation Exposure (ACRE) model calculated a flight-dose profile and an accumulated total dose for each one-way flight of each type of aircraft. ACRE generated 1,895 calculated doses. On the basis of the passenger miles flown on each flight segment and the percent and frequency of flying by the American public, the cumulative and average doses to the crew, flying population, and total U.S. population were calculated. The summary table from the ACRE paper is reproduced as Table 5-1. The paper reported good agreement between a series of in-flight measurements and calculations. It stated:[165]

> The ACRE average estimate resulting from the detailed air travel data is 160 mrem/year/crew member. This dose is less than the radiation guide limit of 170 mrem/year average additional dose above background recommended for the general public, and it is well below the 500 mrem/year maximum for any individual member of the general public. For occupational exposure of radiation workers, the corresponding limit is 5000 mrem/year.
>
> The values for dose equivalent from commercial flying derived here are 0.47 mrem/person/year when averaged over the total U.S. population and 2.8 mrem/person/year for that segment of the total adult U.S. population that traveled by airline at least once during the year. These compare well with the values of 0.48 mrem/person/year and 3.8 mrem/person/year previously reported [by Schaefer in 1972[130]].

Because substantial changes have occurred in the commercial airline industry over the last decade, it is appropriate to re-evaluate the results cited above, which were based, in part, on Civil Aviation Board (CAB) data from the late 1960s and early 1970s. At that time, about 21–25% of the U.S. population surveyed had flown at least once during a 12-mo period. Other CAB surveys estimated that 66% of passengers traveled less than 1,600 km (1,000 miles), and 89.4% less than 3,200 km (2,000 miles). Since the time of the ACRE calculations and the CAB surveys, several changes have occurred in the U.S. commercial air travel industry. Passenger miles grew rather slowly in the early 1980s, but grew at more than 7%/yr between 1983 and 1985. Projected annual growth is 5% into the mid-1990s. In 1984, there were over 343 million domestic passenger enplanements; Figure 1-1 shows that that is expected to reach 500 million by 1995. Commuter-carrier revenue passenger miles, 3.4 billion in 1984, have been increasing rapidly since deregulation and are expected to triple by 1996.[158]

TABLE 5–1 Radiation Doses and Air Travel Statistics Based On Program ACRE[a]

	With 100% Occupancy (where applicable)	With 60% Occupancy
Flights per year	2,991,000	
Flights per day	8,194	
Average number of seats per flight	156	
Seats per day	1,281,000	769,000
Seats per year	468,000,000	281,000,000
Flight crew members per year[b]	16,803	
Cabin crew members per year[b]	22,996	
Seat-kilometers per year	581,000,000,000	349,000,000,000
Flight crew-kilometers per year	9,372,000,000	
Cabin crew-kilometers per year	13,000,000,000	
Seat time, h/yr	736,000,000	442,000,000
Flight crew time, h/yr	12,100,000	
Cabin crew time, h/yr	16,600,000	
Total seat dose, man-mrems/yr	164,300,000	98,580,000
Total flight crew dose, man-mrems/yr	2,650,000	
Total cabin crew dose, man-mrems/yr	3,690,000	
Average flight altitude, km	9.47	
Average flight distance,[c] km	1,084	
Average flight time, h	1.41	
Average dose rate, mrems/h	0.20	
Average dose per flight, mrems	0.28	
Average dose per adult passenger,[d] mrems/yr	2.82	
Average dose per flight crew member, mrems/yr	158	
Average dose per cabin crew member, mrems/yr	160	
Average dose to total U.S. population,[e] mrems/person per year	0.47	

[a] Reprinted with permission from Wallace and Sondhaus.[165]

[b] Assuming a limit of 720 h per full-time equivalent crew member at altitude per year, this number of crew members would be required. "Flight crew" refers to flight-deck crew, and "cabin crew" refers to flight attendants. Crew members flying 480 h/yr—instead of 720—would reduce their doses by a factor of 2/3.

[c] According to FAA estimates, the average flight distance is 1,364 km and the median is 933 km.

[d] Average dose to those who flew in the 12 mo of 1973. Of the total adult population of 140 x 10^6, those who flew in the previous 12 mo were 35 x 10^6, who shared the 98.6 x 10^6 man-rems.

[e] The total yearly dose=(98.6+2.65+3.69) x 10^6=104.9 x 10^6 to passengers, flight deck crew, and cabin crew. This number divided by the total 225 x 10^6 U.S. population gives 0.47 mrem/yr.

Aircraft flights are increasing more slowly than passenger miles because of the trend to the use of larger two-engine jet aircraft in service (see Figure 1-3) and the gradual increase in passenger load factors for domestic flights. The international passenger load factor is expected to remain roughly stable over the next 10 yr (see Figure 1-2).

The FAA Aviation Forecasts (1985–1996) indicate a strong recovery in the domestic aviation industry in 1984 and 1985, after a 4-yr period of operating losses. Furthermore, the composition of the aircraft fleet has changed. Planes are being certified to fly up to 14 km (46,000 ft), where the radiation dose rate is about twice that at 10 km (32,800 ft). The jet aircraft fleet will increase primarily with two-engine wide-and narrow-body planes. This will reduce the number of cockpit flight crew members.

The implication of these changes for the expected radiation dose to the crew will depend on changes in work practices. If increased flights and passenger trips result in increased employment by the airlines, the total radiation dose to the crew will increase, even if the dose to the average crew member does not.

Bramlitt[25] pointed out that the calculations of Wallace and Sondhaus probably underestimated crew radiation dose. Cockpit crew are allowed (by FAA regulations) to fly up to 100 h/m.[48] FAA does not restrict flight-attendant flight time. Some airlines are offering incentives to increase the monthly flight hours of attendants.

Changes in flight altitudes, increases in passenger miles, increases in high-latitude flights, and increases in attendant flight time are expected to increase population and crew radiation exposure. Bramlitt[25, 26] argued that these changes render the Wallace and Sondhaus calculations of cosmic radiation exposure of 160 mrems/yr for flight and cabin crew members inappropriate for 1986. Crew and passengers flying more hours at higher latitudes and altitudes can receive substantially more radiation than 160 and 3 mrems/yr, respectively. By 1995, commercial flying might be expected to increase the integrated cosmic radiation exposure to 1 mrem/person per year when averaged over the total U.S. population. The average exposure of the traveling segment of the U.S. population should stay at about 3 mrems/person per year, unless there is a shift to longer and more frequent flights per person.

Radiation Exposure in Aircraft and from Other Natural Sources

Natural background constitutes the greatest source of ionizing radiation. The exposure is not uniform; such factors as altitude, geologic features, and living structures result in variations. The U.S. population is receiving genetically significant dose-equivalent radiation from natural background that ranges from 40 to 180 mrems/yr (see Figure 5-6). Oakley[112] calculated the population exposure to cosmic and natural terrestrial radiation for the U.S. population on the basis of 1960 census data. He took into account the geographic distribution of population and the altitude, and he extrapolated the effects of terrestrial radiation from the Atomic Energy Commission-sponsored Aerial Radiological Measurement Surveys conducted from 1958 to 1963. Averaged across the U.S. population, a person receives 44 mrems/yr from cosmic radiation and 40 mrems/yr from terrestrial radiation.

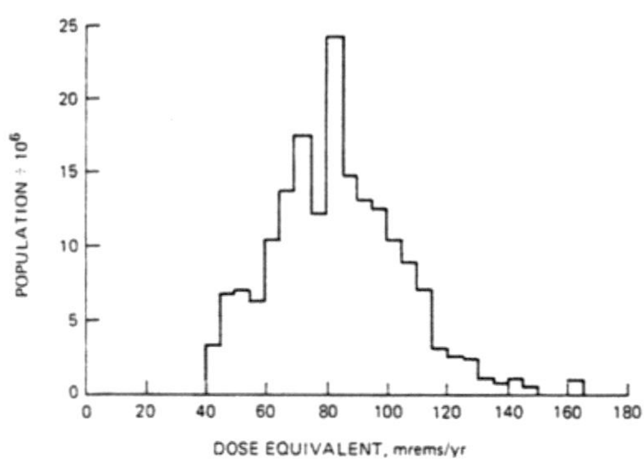

FIGURE 5-6
Population distribution vs. dose equivalent from terrestrial and cosmic radiation. Reprinted from Oakley.[112]

In the mid-1970s, it was generally recognized that terrestrial radiation might be underestimated.[178] Although structural features, such as homes and other buildings, offer some shielding from cosmic radiation (5–20%), structures can increase exposure to natural radiation by leading to accumulation of radon and its decay products indoors. Single-family residences might have a concentration of randon and radon decay products 10 times that outdoors, or even more.

To calculate radiation doses due to flying, one can assume rates of 0.3–0.4 mrem/h at 36,000 ft (11 km) and 0.6–0.8 mrem/h at 45,000 ft (13.7 km). Thus, a passenger or crew member would have to be at these altitudes for only about 100–300 h to receive a dose of ionizing radiation equivalent to that from natural background in a year at sea level.

Groups at Increased Risk of Health Effects

There are approximately 100,000 commercial-aviation crew members in the United States. In addition, about 28% of the U.S. population flies at least once a year. For the vast majority of airline passengers, the additional equivalent radiation dose from flying is less than 3 mrems/yr. A crew member routinely flying 70–83 h/mo can receive a substantial additional dose. Depending on altitude and latitude of routes flown, a crew member might receive up to 1,000 mrems/yr from flying.

Both the National Council on Radiation Protection and Measurements and the International Commission on Radiological Protection recommend that exposure of the fetus during the entire gestation period from occupational exposures of the expectant mother not exceed 0.5 rem.[68 108]

Stewart and co-workers,[144–146] MacMahon,[95] and MacMahon and Hutchison[96] have determined that fetuses are at high risk. They showed that all types of childhood cancer and leukemia are doubled by even extremely small doses of radiation. More specifically, Stewart and Kneale[144] indicated that 1.5 rads from x rays taken in the latter half of pregnancy doubled the frequency of leukemia in children. However, if x rays were taken during the first trimester of pregnancy, only 0.3 rad was needed to double the incidence of cancer in the first 10 yr of life.

Pregnant flight attendants might receive radiation exposure in excess of 500 mrems over the duration of their pregnancy if they fly full-time (70–85 h/mo) on high-altitude flights. Airlines should investigate the policy options for informing female flight attendants about the possible risk involved in flying during pregnancy. In light of the Pregnancy Discrimination Act of 1978,[116] the issues of employees' rights, the rights of fetuses, and airline-industry liability must be addressed in a comprehensive formulation of public and private policy on this matter. Now that the issue of increased radiation exposure among airline employees has been raised by Bramlitt and in this report, FAA and the Environmental Protection Agency, responsible for radiation-protection guidance for occupational exposure, should investigate the in-flight cosmic radiation exposures of crew members. Of particular concern are increased exposures during solar flares. Bramlitt[25] reported that 5 yr of continuous satellite monitoring by the National Oceanic and Atmospheric Administration had shown an average of seven enhanced solar flares per year; one per year is enough to increase the neutron flux on the ground. The time from detection to maximal activity is 19 h. At 40,000 ft (12.2 km), flares can increase the cosmic radiation dose rate from 0.7 mrem/h to 200 mrems/h. Rare events can increase the rate to 2,000 mrems/h.

GROUND FUMES

While waiting for a plane to depart or arrive and while sitting in a taxiing plane, passengers can be exposed to substances emitted by aircraft engines and the engines of maintenance vehicles. There is relatively little information on actual exposures in aircraft during these periods, but some information on potential exposure to various substances can be obtained from a review of aircraft engine emission.

Aircraft jet engines emit a variety of potentially toxic substances, [92 93 118 136] including carbon monoxide, oxides of nitrogen, hydrocarbons, aldehydes (especially formaldehyde), particles, and polynuclear aromatic compounds. The rate of emission varies with the operation of the aircraft. When idling, engines are less efficient and might emit higher concentrations of some pollutants (e.g., carbon monoxide and hydrocarbons, but not oxides of nitrogen). Aircraft idling and taxiing are major sources of airport air pollution,[92] and idling time is often limited to meet local air pollution criteria.

A study in 1970 evaluated aircraft engine emission as a source of air pollution at Los Angeles International Airport chiefly by monitoring carbon monoxide and particle concentrations in and around the airport. [92] It also included monitoring for carbon monoxide in aircraft on the runway or at the gate. This study demonstrated that carbon monoxide concentrations in the cabin paralleled those outside the aircraft. Cabin concentrations were highest (approximately 10–15 ppm) when the airplane was at the gate loading passengers; that reflected the higher concentrations of carbon monoxide (and particles) in that area of the airport. In general, the study found the highest concentrations of particles and carbon monoxide in or around the airport to be near the passenger terminals, where air and ground traffic was greatest.

Although ground fumes from jet engine exhaust contain substances that can cause respiratory irritation and other health effects, there is little available information from monitoring that indicates the exposure of cabin occupants to these substances.

ENVIRONMENTAL TOBACCO SMOKE

The air contaminant in an aircraft cabin that is most apparent to the passengers and crew is cigarette smoke. Cigarette-smoking contributes to environmental tobacco smoke (ETS) in four ways: it contributes smoke from the smoldering ends of cigarettes (sidestream smoke), smoke that escapes during puff-drawing from the burning cone, vapor that escapes through the paper of the cigarette, and smoke exhaled by smokers. Secondary reactions in these diluted smokes alter their physical and chemical characteristics. ETS is a complex mixture of gases and particles.[161]

The proportion of passengers who are current cigarette smokers can be estimated from statistics that describe the passenger population (Chapter 1) and the distribution of smokers in the general American population. Some 54% of passengers are male and 46% female;[51] 37% of American males and 29% of females currently smoke.[151] Therefore, the proportion of passengers who currently smoke is (0.37)(0.54) + (0.29)(0.46) = 32.3%. This agrees with the observation that somewhat less than one-third of passengers request seats in the smoking section.

In 1970 and 1971, before establishing smoking restrictions on aircraft, FAA and the U.S. Public Health Service conducted a questionnaire survey of 20 military flights and 14 domestic civilian flights in conjunction with ambient air assessments (described in some detail later in this report).[160] In that study, 31% of the domestic passengers smoked on the flights (an average of 2 cigarettes each), and 52% of the military passengers smoked on the flights (an average of 8 cigarettes each). In 1961, Halfpenny and Starrett[58] found that smokers average 1.25 cigarettes/h on 2-h flights and that 51% of passengers smoke on aircraft. Both these studies were conducted before the smoking restrictions on aircraft. The current estimates of smoking rates are 2.1 and 2.2 cigarettes/h.[32] [148] The above numbers are estimated averages, and the actual smoking rates on aircraft are highly variable, as illustrated in Figure 5–7.

Some aspects of cigarette-smoking on airplanes are peculiar to that situation. In public places generally, it might be expected to find one person in nine smoking at any given time. However, on aircraft, smokers are seated together, and smoking might be heaviest after the "no smoking" light is turned off and after a meal is consumed. This pattern of smoking results in higher transient concentrations of cigarette smoke than occur in other public places where smoking is permitted. High transient concentrations occur not only in the smoking section, but also in other parts of the cabin.

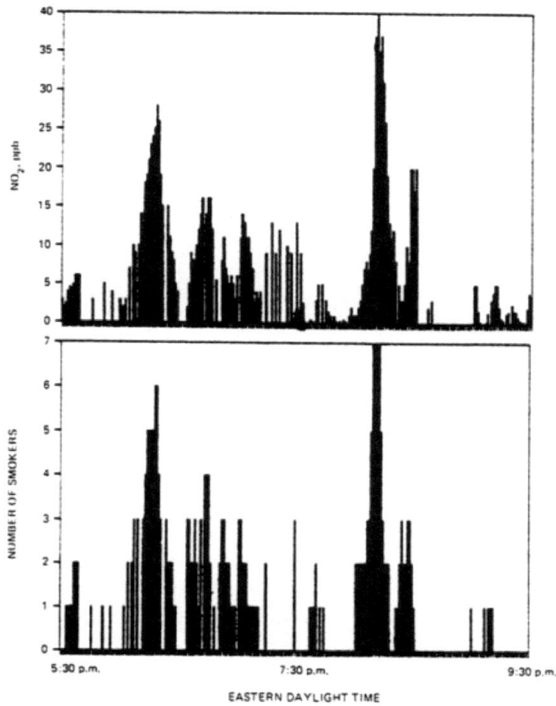

Figure 5-7
Top, No_2 concentration vs. time during flight from Boston to Denver. Bottom, number of cigarettes smokers on same flight. NO_2 measured approximately once a minute; smokers (passengers with lighted cigarettes) counted approximately 15 s after NO_2 measurement. Data from D. H. Stedman (personal communication, 1985).

Aircraft Ventilation and Smoke Concentrations

Table 5-2 is a partial list of compounds in cigarette smoke. Many of these are more heavily concentrated in the sidestream. The sidestream-mainstream ratios presented in the table were measured under standardized laboratory conditions. In the cabin air, the relationship among the constituents in ETS varies with the brand of cigarette, smoking behavior, and environmental conditions (e.g., humidity and air mixing). Many of the chemical components of ETS are known to be toxic (e.g., acrolein and carbon monoxide) or carcinogenic (e.g., N-nitrosodiethylamine and benzo[a]pyrene) in humans or animals.

As discussed in Chapter 7, measurements of carbon monoxide, nitrogen oxides, respirable suspended particles (RSP), and light-scattering are all used as surrogates to detect ETS. It is reasonable to assume that the harmful components are proportional to the measured gas phase and particle phase of ETS.

If cigarette-smoking on aircraft were at a constant rate, a steady-state concentration of smoke would be achieved after 5–10 min under typical ventilation rates—about twice the typical air-exchange time (see Chapter 2). On aircraft without recirculated air, the steady-state density will be approximately proportional to the product of the rate of smoke production and inversely proportional to the flow of outside air.

Other factors also affect ETS concentrations, such as deposition on surfaces, especially fabrics. Under normal conditions, the rate of chemical and physical removal in the cabin is much less than the rate of removal by ventilation. The ETS concentration can be decreased by decreasing the source (i.e., the number of cigarettes smoked) or by increasing the rate of flow of outside air (i.e., decreasing the air-exchange time). Decreasing the size of the smoking section might increase the concentration of ETS in that section, if the same number of smokers are concentrated in fewer seats.

The above relationships neglect the recirculation patterns common on modern aircraft. In most aircraft, there is no physical barrier between the smoking and nonsmoking sections. Consequently, there will be some mixing between sections. In some wide-body jets, air is recirculated to the zone from which it is taken. In this design, if a zone is designated smoking or nonsmoking, recirculation should not affect mixing in other areas of the aircraft. Recirculation patterns in which air is mixed throughout the whole aircraft distribute gaseous smoke products and submicrometer particles throughout the aircraft. The filter systems described in Chapter 2 should be adequate to remove micrometer-sized particles and a portion of the submicrometer smoke particles. However, ETS vapors would not be removed. Furthermore, if the optional charcoal absorption beds are installed and maintained, gaseous contamination will be substantially reduced. However, efficiency of charcoal absorption varies with compounds, water vapor, flow rate and time. These complexities could be taken into account in the model described in Appendix A.

TABLE 5-2 Distribution of Compounds in Nonfilter-Cigarette Undiluted Mainstream and Diluted Sidestream Smokea

Compound	Total Emission in Mainstream Smoke, μg/cigarette	Sidestream-to-Mainstream Total Emission Ratio
Vapor phase:		
Carbon monoxide	10,000–23,000	2.5:1–4.7:1
Carbon dioxide	20,000–40,000	8:1–11:1
Carbonyl sulfide	18–42	0.03:1–0.13:1
Benzene	12–48	10:1
Toluene	160	6:1
Formaldehyde	70–100	0.1:1–50:1
Acrolein	60–100	8:1–15:1
Acetone	100–250	2:1–5:1
Pyridine	16–40	6.5:1–20:1
3-Methylpyridine	12–36	3:1–13:1
3-Vinylpyridine	11–30	20:1–40:1
Hydrogen cyanide	400–500	0.1:1–0.25:1
Hydrazine	0.032	3:1
Ammonia	50–130	40:1–170:1
Methylamine	11.5–28.7	4.2:1–6.4:1
Dimethylamine	7.8–10	3.7:1–5.1:1
Nitrogen oxide	100–600	4:1–10:1
N-Nitrosodimethylamine	0.01–0.04	20:1–100:1
N-Nitrosopyrrolidine	0.006–0.03	6:1–30:1
Formic acid	210–490	1.4:1–1.6:1
Acetic acid	330–810	1.9:1–3.6:1

Compound	Total Emission in Mainstream Smoke, μg/ cigarette	Sidestream-to-Mainstream Total Emission Ratio
Particulate phase:		
Particulate matter	15,000–40,000	1.3:1–1.9:1
Nicotine	1,000–2,500	2.6:1–3.3:1
Anatabine	2–20	<0.1:1–0.5:1
Phenol	60–140	1.6:1–3.0:1
Catechol	100–360	0.6:1–0.9:1
Hydroquinone	110–300	0.7:1–0.9:1
Aniline	0.36	30:1
2-Toluidine	0.16	19:1
2-Naphthylamine	0.0017	30:1
4-Aminobiphenyl	0.0046	31:1
Benz[a]anthracene	0.02–0.07	2:1–4:1
Benzo[a]pyrene	0.02–0.04	2.5:1–3.5:1
Cholesterol	22	0.9:1
γ-Butyrolactone	10–22	3.6:1–5.0:1
Quinoline	0.5–2	8:1–11:1
Harman	1.7–3.1	0.7:1–1.7:1
N'-Nitrosonornicotine	0.2–3	0.5:1–3:1
NNK[b]	0.1–1	1:1–4:1
N-Nitrosodiethanolamine	0.02–0.07	1.2:1
Cadmium	0.1	7.2:1
Nickel	0.02–0.08	13:1–30:1
Zinc	0.06	6.7:1
Polonium-210	0.04–0.1 pCi	1.0:1–4.0:1
Benzoic acid	14–28	0.67:1–0.95:1
Lactic acid	63–174	0.5:1–0.7:1
Glycolin acid	37–126	0.6:1–0.95:1
Succinic acid	110–140	0.43:1–0.62:1

[a] Total emissions are given for fresh, undiluted mainstream smoke generated by a smoking machine under conditions of 1 puff/min of 2-s duration and 35-ml volume, i.e. 10 puffs/cigarette. Sidestream values are given for smoke collected with an airflow of 25 ml/s, which is passed over the burning cone. Compiled by D. Hoffmann (personal communication, 1986) from Elliott and Rowe,[45] Hoffman et al.,[64] Klus and Kuhn,[78] Sakuma et al.,[124–126] and Schmeltz et al.[134]

[b] 4-(N-Methyl-N-nitrosamino)-1-(3-pyridyl)-1-butanone.

Concentrations of ETS Constituents Measured on Aircraft

Aircraft air quality has not been a subject of systematic investigation by independent researchers. Various airlines have conducted their own studies of airborne contaminants. Several airlines—such as Air France,[164] United Airlines,[8] and Lufthansa German Airlines[94]—have conducted tests, and some Committee members have conducted a few "measurements of opportunity." That is, the measurements have not been conducted under experimental situations or have not been conducted systematically for a variety of aircraft. As discussed in Chapter 7, isolated measurements are likely to be highly variable, even if made with accurate instruments.

The distribution of smoke in the aircraft cabin is not uniform, but rather exhibits spatial and temporal variability. The concentration measured in any area would depend on location of the sampler in relation to the smoke source and the ventilation in that area.

In 1970 and 1971, in one of the earliest studies, FAA and the U.S. Public Health Service[160] measured carbon monoxide, aromatic hydrocarbons, aldehydes, ketones, and total particulate mass on 20 military flights and 14 domestic civilian flights. These studies were done before smokers were segregated in the aircraft cabin, so their relevance to present conditions is not clear.

Data from more recent studies are listed in Table 5-3. Lufthansa[94] provided material that contained useful information about relative humidity; however, because the instruments used for measuring contaminants had limits of detection above the expected values, these data are not included in the table.

Members of the Committee have used portable instruments to measure ETS concentrations on commercial flights. These measurements were not accompanied by detailed documentation of ventilation or numbers of people smoking. They are included here only to illustrate further the concentrations that could be encountered on aircraft. A hand-held nephelometer (see DC-10 flight data from Spengler in Figure 5-3) and piezoelectric balance (see B-747 flight data) were used to measure mass concentration of suspended particulate

TABLE 5-3 Examples of Measurements of Pollutants on Airliners

Source of Measurement	Aircraft	Constituent Measured	Concentration
D. H. Stedman (personal communication, 1985)	B-727-200	NO_2	0–40 ppb
FAA and USPHS[160]	Several, 1970–1971	CO	Max., 5 ppm
		RSP	Avg., 140 μg/m^3; peak, 1,200 μg/m^3
United Airlines[8]	B-747	CO	Max., 3 ppm
		RSP	60–320 μg/m^3
	DC-10	CO	Max., 5 ppm
		RSP	19–400 μg/m^3
	DC-8-61	CO	Max., 5 ppm
		RSP	70–260 μg/m^3
	B-727	CO	Max., 5 ppm
		RSP	40–140 μg/m^3
	B-737	CO	Max., 5 ppm
		RSP	80–200 μg/m^3
Air France[164]	B-747	CO	Max., 5 ppm
J. Spengler (personal communication, 1986)	B-747	RSP	10–50 μg/m^3 nonsmoking
		RSP	50–500 μg/m^3 in smoking section; peak, 1,000 μg/m^3
	DC-10[a]	RSP	10–40 μg/m^3 in nonsmoking aft cabin with no cigarette odor
		RSP	100±20 μg/m^3 in nonsmoking forward cabin with cigarette odor
		RSP	300±200 μg/m^3 in smoking section; peak, 750 μg/m^3
		CO_2	550–1,200 ppm

[a] Load factor, 40–60%.

matter. The nephelometer responds optimally to particles in the submicrometer range. The RSP concentrations were about 10–50 μg/m^3 in the two-thirds-filled nonsmoking section of a wide-body airliner, about 100 μg/m^3 at the front of the smoking section, and over 500 μg/m^3 in the rear of the smoking section near the lavatories. Occasional readings exceeded 1,000 μg/m^3. Similar concentrations were recorded on a DC-10 over six segments of a round-trip flight between Boston and Anchorage. Load factors were between 40 and 60%.

Standards for Other Environments

There are no federal standards for ETS in any environment, although smoking has been prohibited in many public buildings by municipal and state ordinances. The occupational and ambient standards for carbon monoxide and particulate matter that are often applied to ETS do not take into account the other toxic materials present in ETS, which contains measurable concentrations of several known carcinogens and cocarcinogens.

The national ambient air quality standards for carbon monoxide and total suspended particles (TSP) are shown in Table I-1 (in the introduction). An additional indoor air standard for particle density in office buildings is the Japanese standard of 150 μg/m^3.[9] For carbon monoxide, the EPA and ASHRAE standards of total 1-h concentration of 35 ppm and 8-h concentration of 9 ppm appear unlikely to be violated in typical airliner cabins. However, the TSP standard is a particle-mass standard designed mainly for protection from pollutants like fly-ash, and not designed to take into account the toxicity or size distribution of ETS. The TSP standards (150–260 μg/m^3 for 24 h) also do not take account of particle size. That is, the TSP standard deals with only total mass, which usually is dominated by larger particles of a size that ordinarily cannot enter the lungs during breathing. However, respirable particles have little mass. A standard that would be specific to RSP is likely to be considerably lower than a comparable TSP standard, because RSP contributes little to the TSP mass. Because aircraft cabin RSP concentrations of 250 μg/m^3 are not unusual, it is apparent that a majority of the air quality measurements given in Table 5-3 would violate the Japanese standard for particle density and, in many cases, the less stringent EPA 24-h standard for TSP.

Ventilation standards for smoking areas in other public places are designed to produce acceptable air in which there are no known contaminants at harmful concentrations and with which a substantial majority (80%) of the occupants do not express dissatisfaction. These standards led to the ASHRAE suggestion of ventilation at 20–50 cfm/person for a variety of settings where smoking is allowed.[4] The maximal flow ventilation distribution in 1985, shown in Figure 2-6, indicates that about 80% of the flights had airflow of less than 40 cfm/passenger. By the above guidelines, it is apparent that aircraft ventilation would not meet the criterion of acceptability to at least 80% of nonsmokers if the nonsmokers were forced to work in, traverse, or wait in an active smoking section.

Exposure to ETS on Airliners

According to a National Research Council report (see National Research Council, Committee on Indoor Pollutants,[110] p. 8), "public policy should clearly articulate that involuntary exposure to tobacco smoke has adverse health effects and ought to be minimized or avoided where possible." Several different groups of people are characterized by different kinds of exposure to ETS.

On commercial aircraft, the people with the greatest exposure are the cabin crew, who are exposed to ETS regularly. In some aircraft, the galley is in the smoking section, so cabin crew are exposed to ETS at the same concentrations and for the same durations as passengers in the smoking section. Thus, cabin crew, including pregnant flight attendants, are likely to be exposed to ETS at high concentrations. Although policies vary among airlines, some attendants are permitted to fly (with their doctors' permission) up to the twenty-eighth week of pregnancy.

Passengers will not be exposed daily. However, nonsmoking passengers in the smoking section, such as spouses and children, will be exposed to the ETS. Passengers in the few nonsmoking rows adjacent to the smoking section are likely to be exposed to the next highest transient concentrations, because of air motion and ETS diffusion from the smoking section. In aircraft without air recirculation, passengers well into the nonsmoking sections, flight crew members, and cabin crew members whose duties do not take them into the smoking sections are relatively unexposed.

The nature of exposure to ETS and its composition is complicated by the fact that all aircraft now in production have some form of recirculation system. The complexity arises because of differences in ventilation equipment between aircraft and differences in operating procedures that change the proportion of outside to recirculated air. In addition, there is usually a filter that removes some particulate matter; however, the passage of gases through the filter is usually unimpeded. Thus, the composition of ETS after it passes through filters has not been characterized for the full range of filters that might be found on an airplane. Cain et al.[33] demonstrated in chamber studies that nonsmokers report dissatisfaction with and irritation by cigarette-generated smoke, even when the smoke is filtered with an electrostatic precipitator. This was true when smoke concentrations were low, as determined by measurement of surrogate carbon monoxide concentrations at 5 ppm and even as low as 1 ppm. Filtration of 80% of particles with Cambridge filters, which are currently in use on aircraft that have recirculation systems, has reduced irritation substantially[111 168] (see Chapter 2).

Health Effects in Airplanes

The irritant properties of cigarette smoke have given rise to complaints about the quality of aircraft environments. Irritation affects general health and welfare and thus affects performance of the crew. Records of passenger or flight attendant complaints compiled by the Association of Flight Attendants[15] listed "smoky" as a complaint in 73 of 297 air quality complaints; the cause was listed in only 113 of the 297 cases. In a 1980 questionnaire study of 1,961 Scandinavian Airlines System (SAS) cabin attendants, only 4% were not at all bothered by smoky air, whereas 69% were "bothered to a great extent." The data are shown in Table 5-4.

TABLE 5-4 Results of 1980 Questionnaire Survey of SAS Cabin Attendantsa

	Attendants Bothered by Factors Listed Below, %		
Subject of Complaint	Not at All	To a Certain Extent	To a Great Extent
Noise	13	53	34
Cold	29	56	15
Cabin temperature variation	32	55	13
Heat	43	49	8
Variation in cabin pressure	36	51	13
Drafts	27	47	26
Static electricity	44	45	11
Dry air	10	31	59
Turbulence	22	60	17
Dust	62	31	7
Smoky air	4	26	69
Odors	26	61	13
Pungent smells	59	34	7

[a] Data from Ostberg and Mills-Orring.[113]

In one study of six nonsmoking flight attendants, increases in blood nicotine and urinary cotinine (a nicotine metabolite) were observed after flights of 8 h.[49] However, Duncan and Greaney[44] found no increase in carbon monoxide in exhaled breath of 16 flight attendants after 10 h of flying (Los Angeles to Honolulu and back).

Health Effects in Other Environments

Given the limited number of studies of exposure to ETS in aircraft, evidence of adverse health effects necessarily is inferred from studies in other environments. These include studies of chronic exposure, relevant to the cabin crew, and studies of acute effects of exposure, relevant to the passengers.

The possible health effects of ETS on nonsmokers cannot simply be extrapolated from the health experience of active smokers, for the following reasons. First, smokers and nonsmokers differ in exposure and deposition of smoke particles in the lung.[60] Particles in mainstream smoke (which the smoker takes in from the cigarette) become much larger than those in sidestream smoke, because they are more highly concentrated and agglomerate in the respiratory tract. Because of lung aerodynamics, larger particles (±1 μm) tend to be deposited in the bronchus, whereas smaller particles (<1 μm) can be carried deeper into the lung and deposited in the smaller tubules and alveolar sacs.[69] The extent to which ETS particles are hygroscopic and increase in size will affect the deposition pattern. However, 89% of the inhaled sidestream smoke particles are exhaled.[61] The membranes of various regions of the lung differ substantially, e.g., in thickness and presence of cilia. Second, the deposition of smoke particles in the lung is also affected by the breathing patterns of the individual; some smokers inhale smoke more deeply than nonsmokers. The 1982 EPA report on particles and sulfur oxides[154] discusses deposition of particles for nose breathing, compared with mouth breathing. The deposition curve peak shifts downward from 3.5-to about 2.5-μm diameters. The peak is much less pronounced (about 25%, compared with 50% for mouth breathing), with a nearly constant pulmonary deposition of about 20% for all sizes between 0.1 and 4 μm.

In a 1985 Gallup poll,[3] over 85% of nonsmokers and 60% of smokers felt that smoking should be restricted in the workplace and that, in the presence of nonsmokers, smokers should curtail their smoking. Speer[140] interviewed nonallergic people regarding their subjective reactions to cigarette smoke. They reported eye irritation, headaches, nasal symptoms, and coughing. Unacceptable odor is often the first complaint of people exposed to ETS. Nonsmokers are more likely to find ETS odor unacceptable than smokers.[32] Cain et al.[32] systematically varied environmental conditions—including temperature, humidity, and ventilation rates—to determine the intensity of ETS-associated odor for visitors to and occupants of an experimental chamber. They found that odor sensitivity to ETS increased as relative humidity was increased from 50% to 75%. Tobacco smoke odor does not decay rapidly.[36]

The odor characteristics of ETS in the airliner cabin need to be studied, to determine how low humidity and other environmental conditions affect discomfort of these types. Cain et al.[32] determined that, with as few as 10% of the occupants in a space smoking at any time, a ventilation rate of 5.3 cfm/occupant was required to make the air acceptable to at least 80% of the occupants, especially nonsmokers, who are more sensitive to the odor than smokers.[32] Kerka and Humphreys[77] demonstrated that, although perceived tobacco smoke odor is reduced over time owing to olfactory adaptation while the person stays in the chamber, the degree of sensory irritation increases. Both odor and irritation are perceived to be more intense at lower humidities (30% vs. 65%) (Figure 5-8). The Committee could find no information on studies done at relative humidities below 10%, which are typical of aircraft. The eye is the most readily affected site of irritation. Weber and colleagues have studied the effects of ETS on the eyes extensively.[168–172] There are no data on the combined effects of ETS, low humidity, and photochemical oxidants (including ozone, formaldehyde, and acrolein) on the eye; contact-lens wearers in particular should be studied in this environment.

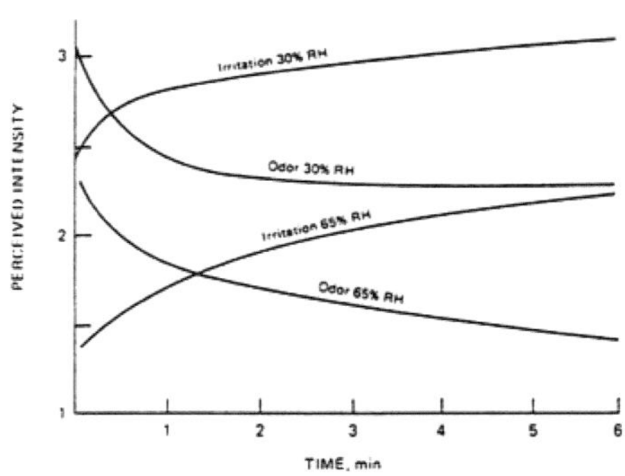

Figure 5-8
Relationship of relative humidity to odor and irritation during continuous short-term exposure to cigarette smoke generated in a chamber. Ventilation, 14 cfm/cigarette; ambient temperature, 25°C. Adapted from Kerka and Humphreys.[77]

Other health effects of ETS have been studied in chronically-exposed nonsmokers compared with unexposed nonsmokers. Chronic exposure studies are more relevant to flight attendants who are chronically exposed occupationally than to passengers who are not otherwise chronically exposed. Extrapolation from these studies to the experience of persons in an aircraft cabin is not straightforward, but the studies do indicate the range of possible health effects to consider.

There have been several studies of lung cancer risk in nonsmoking spouses of smokers. In 1981, studies in Japan[62 63] and Greece[150] showed that women with smoking husbands had a statistically significantly higher risk of lung cancer than other women and that the risk increased with the number of cigarettes that their husbands smoked. Since then, several investigators have examined this association with case-control and prospective studies.[35 41 52 55 62 63 76 80 82 127–129 150 177] Because of sample sizes, most of the observed differences are not statistically significant. However, of the 15 studies that separated nonsmokers from smokers, most found an increase in risk associated with chronic exposure to ETS. A positive association of lung cancer with ETS exposure is biologically plausible (ETS does contain toxic and carcinogenic chemicals), and results are consistent between studies and across study designs and cultural settings. Therefore, the Committee concludes that there is a positive association between lung cancer and chronic exposure to ETS. Exposure values in these studies were developed from questionnaire data that indicated that the nonsmoking subjects were chronically and regularly exposed to ETS at home. The Committee on Passive Smoking of the National Research Council is currently reviewing the available published literature on the health risks associated with ETS exposure and will prepare a report.

To evaluate the relevance of occupational exposures to ETS for the cabin crew in aircraft, we assumed that a flight attendant worked 800 h/yr in the smoking section of an airplane, where the average concentration of total particles might be 250 $\mu g/m^3$. Assuming a breathing rate consistent with modest exercise —i.e., 15 L/min—under these circumstances the integrated exposure to ETS would be 1.8×10^5 µg/yr. Spengler et al.[142] reported that a home with a 1-pack/d smoker is likely to have particle concentration at least 20 μg/m^3 higher than a home without smokers. But if a nonsmoker lives with a 1.5-pack/d smoker, sharing approximately 70% of his or her time at home and breathing at a resting rate of 10 L/min, the nonsmoker would have an integrated exposure of 1.1 x 10^5 μg/yr. Thus, it is likely that a flight attendant working full-time is receiving an integrated exposure to ETS approximately equal to that associated with living with a 1.0-pack/d smoker.

Most studies of the effects of ETS on the lung of adults have investigated pulmonary function changes that might indicate early disease. Many studies of chronic exposure of children indicate that the prevalences of respiratory symptoms and illness are increased and pulmonary function can be descreased.[161] However, there have been relatively few studies in adults. White and Froeb[174] reported that nonsmoking healthy adults exposed to tobacco smoke at work had lower forced expiratory flow rates than nonsmokers not so exposed. However, these results have been questioned by several investigators.[1 16 67 87] Kentner et al., [76] in another study of the effect of smoking in the work environment, found no significant change in the results of any pulmonary function test among working adults. There have been a number of studies of changes in pulmonary function of adults in relation to exposure to smoke in the home environment. Comstock et al.[39] and Brunekreef et al.[29] detected decreases in pulmonary function, although the decreases were insignificant. Kauffmann et al.,[74 75] in two studies, detected significant decreases in standardized forced expiratory volume, especially in women over age 40. Thus, it is difficult to draw any firm conclusion on the use of forced expiratory rates in determining health effects of ETS. It is unclear whether the investigators were treating the observed changes in pulmonary function as representative of a health effect of long or short exposure to ETS and whether the changes were symptomatic of pulmonary disease. There were no assessments of ambient conditions or biochemical measures of components of tobacco smoke in most studies. Therefore, it is difficult to extrapolate the results of these studies to potential health effects in passengers and crew members in airliner cabins. Some chemicals in tobacco smoke, however, are known to cause mucociliary stasis, toxicity to alveolar macrophages, increased permeability of the mucosal barrier, and changes in immunoglobulins.[38 47]

There are few data on the potential effects of ETS on cardiovascular disease. Investigations have centered on carbon monoxide and nicotine, because of their known effects on the oxygen-carrying capacity of blood and on the sympathoadrenal system.[162] A recent comprehensive review of laboratory and clinical data on animals and humans with respect to nicotine and carbon monoxide uptake in passive smoking and its potential effect on the cardiovascular system[133] concluded that passive smoking should have no cardiovascular effects in humans. However, the reports of cardiovascular complications in previously normal people exposed to ETS raise the possibility of deleterious effects associated with exposure.[53 55 62 63]

In summary, the cabin crew are chronically exposed to substantial ETS concentrations. The total exposures might approach those experienced by spouses of smokers. Therefore, the health effects assessed in spouses of smokers could be relevant for the cabin crew. The cabin crew and asymptomatic passengers are subject to acute health effects of exposure to ETS. Furthermore, given the low relative humidity and other environmental conditions of the cabin, such as high ozone concentration, the irritation and discomfort effects are likely to be important for occupants of the smoking and nearby sections and to others who need to move through the smoking section.

Groups at Increased Risk

Other persons who might have a different risk of exposure to ETS on aircraft are passengers with pulmonary or cardiovascular diseases. Dahms et al.[42] found that asthmatic patients had a significant linear decrease in pulmonary function after exposure to ETS, with reductions in forced expiratory volume, forced expiratory flow, and forced vital capacity. The etiology of these changes has not been defined clearly, although some suggest that the smoke might increase airway resistance in patients with bronchial asthma, whose pulmonary function was already lower than that of normal subjects. Wiedemann et al.[176] also studied asthmatics. However, his patients were not on medication and had normal or nearly normal lung function. After their exposure to ETS, they found a significant decrease in airway reactivity, as assessed by a methylcholine challenge test. Shephard et al.,[137] however, found decreases in pulmonary function among medicated asthmatic patients when they were exposed to ETS, but the changes were not significant.

It has been suggested but never proved, that patients with angina are at increased risk of recurrence in the presence of ETS. Aronow et al.[11 14] reported an exposure-related subjective outcome, angina, in persons with severely compromised cardiovascular systems—patients who had previously suffered from angina pectoris. Similar findings of early-onset angina were observed when patients were exposed to carbon monoxide at concentrations characteristic of those noted during the ETS exposure experiments.[10 12 13] The validity of these findings has been questioned,[155] and the studies are currently being repeated by both the National Institute of Environmental Health Studies and the Health Effects Institute.

Other groups that might be at increased risk due to exposure to ETS are people with various chronic pulmonary diseases, including chronic obstructive pulmonary disease, emphysema, alpha-1-antitrypsin deficiency, and cystic fibrosis. However, the effects of ETS on these people have not been studied.

Prevention of Exposure to ETS

Occupational exposure of flight attendants to ETS could be limited by configuring aircraft without any work stations (galleys) in the smoking section. Exposure of nonsmoking passengers would be lessened if access to lavatories did not require passage through the smoking section. Total isolation of smokers and their air is possible, but would be a major engineering task whose cost would presumably be borne by the flying public through higher ticket costs.

Light-weight, high-performance, economical filter systems that effectively remove gases and particles from ETS could eliminate many of the problems of and objections to onboard smoking. Such systems that are compatible with requirements for installation on airplanes have not yet been developed.

The ultimate prevention of exposure will be achieved only when there is no smoking on aircraft. However, it has been argued that the acute withdrawal from compounds in tobacco smoke would be accompanied by symptoms that are both physical and psychologic in origin and that heavy smokers might find it difficult to endure the stress of long airplane flights without smoking. Murray and Lawrence[106] have described the state of knowledge regarding withdrawal symptoms. The weight of evidence does not support the view that unpleasant physical and psychologic effects necessarily follow abstinence from smoking. Weight gain, craving for cigarettes, etc., are highly idiosyncratic and do not occur with high frequency in smokers who are temporarily required to cease smoking. Those who suffer withdrawal symptoms should discuss with a doctor the use of nicotine substitutes to alleviate the discomfort.

The Civil Aeronautics Board (CAB) suggested in 1981, and later withdrew, a ban on smoking on flights lasting less than 2 h.[153] The ban was withdrawn because

> it would be particularly troublesome on multisegment flights where smoking would be permitted at some times, and then prohibited during other parts of the trip. In addition, such a ban would require arbitrary line drawing. It might create a perverse incentive for some carriers to rearrange their schedules to evade it. We conclude on balance that the limited benefits are outweighed by the difficulties associated with the proposal.

In the judgment of the Committee, the potential health effects of passive smoking are of more concern than effects of withdrawal, and more people are at risk. We believe that CAB's suggested 2-h limit did not provide adequate protection to passengers on longer flights. Therefore, the ban on smoking should be extended to all domestic flights. Limiting the ban on smoking to domestic flights would allow smokers the option of taking planes with stopovers that would enable them to smoke in smoking areas of the airport. The use of nicotine substitutes could discourage surreptitious smoking, as could strict enforcement of no smoking rules with the threat of fines. The hazard of in-flight fires resulting from surreptitious smoking in lavatories can be reduced through the use of nonremovable smoke detectors.

After a period of adjustment and with strict enforcement, prohibiting smoking should reduce onboard fire risk, cleaning costs, and costs of replacing and repairing damaged materials. Removing tobacco smoke from the aircraft environment would reduce cabin ventilation requirements, and that would result in additional fuel savings while reducing irritation and health risk.

Summary

The concentration of ETS is directly proportional to the strength of the source (number of active smokers) and inversely proportional to the flow of outside air in a smoking area. Recirculation in various configurations complicates the distribution, but does not fundamentally change the relationship between smoke generation and the steady state of the total mass of contaminants in the cabin. Currently available filters on airplanes only remove particles. Some particles and tars are removed through settling and adsorption onto cabin surfaces.

Carbon monoxide and respirable particulate matter are measured as surrogates for ETS, which is a complex mixture of many components. Peak concentrations of carbon monoxide and RSP of about 5 ppm and 500 $\mu g/m^3$, respectively, are to be expected and have been measured. However, the data supporting these values are sparse, and most have not been subjected to peer review.

The measured values do not violate U.S. ambient or workplace carbon monoxide standards, but do violate a Japanese standard for indoor air quality. Cigarette smoke contains known human and animal carcinogens that would be strictly regulated if the source were something other than tobacco. Ventilation standards that have been set to avoid irritation by ETS in buildings are not met by standard aircraft practices.

The most regularly exposed nonsmoking populations are cabin crew members whose duties require them to spend long periods in the smoking section and passengers who are seated in or near the smoking section.

Health-effects data from other environments do not permit us to present reliable quantitative risk estimates related to the health impact of present concentrations of

ETS on exposed nonsmokers in an aircraft environment. One report that presented a risk-assessment calculation for ETS suggested that a reduction by more than a factor of 10 in present aircraft concentrations would be necessary to bring the risk calculated in that report into the range permitted for regulated toxic environmental contaminants. This degree of change, which might be technologically possible, is likely to be economically unrealistic if smoking were permitted in aircraft.

Patients with severe asthma or angina are at higher risk from exposure to ETS than other exposed people, because of the increased likelihood of acute symptoms.

The Committee considered several ways of reducing ETS in aircraft. Solutions requiring structural or engineering changes—such as increasing ventilation, moving lavatories and galleys, and separating smoking compartments by physical barriers—are not likely to be economically feasible. Increasing ventilation for the smoking zone to be in compliance with ASHRAE guidelines is not technically feasible on existing aircraft. The amount of air that can be extracted from the engines is limited and might not support the high ventilation rates; in addition, the high rates would require ECU redesign, increased distribution ducting, outlet redesign, and control modification. A return to the random distribution of smokers throughout the cabin would decrease the peak concentrations of contaminants, but the Committee feels that this probably would be unacceptable to a majority of the traveling public.

The Committee recommends a ban on smoking on all domestic commercial flights, for four major reasons: to lessen irritation and discomfort to passengers and crew, to reduce potential health hazards to cabin crew associated with ETS, to eliminate the possibility of fires caused by cigarettes (see Chapter 1), and to bring cabin air quality into line with established standards for other closed environments (see discussion on ventilation in Chapter 2 and on specific pollutants in Chapter 5). The ban might have the added benefit of reducing airline maintenance costs for removal of tobacco tars. We note that some habitual smokers might experience nicotine deprivation on flights longer than 3 h. However, in the judgment of the Committee, the potential health effects of passive smoking are of more concern than the discomfort of withdrawal, and more people are at risk.

BIOLOGIC AEROSOLS

Types of Biologic Pollutants Possible in Aircraft

Most biologically derived particles that are known to become airborne could be present in aircraft cabin air. These include viruses, bacteria, actinomycetes, fungal spores and hyphae, arthropod fragments and droppings, and animal and human dander.[30]

Viruses that are known to be infective through the airborne route include rhinoviruses, influenza viruses, coxsackievirus, adenovirus, and measles virus. Disease transmission through the air is known to occur both by droplets and by droplet nuclei, which can be transported over relatively long distances.[79]

A wide variety of bacteria have been isolated from air. Those which have caused disease through airborne carriage include streptococci, mycobacteria, staphylococci, legionellae, pseudomonads, and klebsiellae.[79] Actinomycetes (so-called filamentous bacteria) that cause invasive disease are rarely isolated from air, but thermophilic (heat-loving) actinomycetes that have been implicated in hypersensitivity pneumonitis always produce disease through the airborne route. These include one or more species in the following groups: Thermoactinomyces, Thermomonospora , Saccharomonospora, and Micropolyspora.[20]

Most fungi produce spores and often hyphal fragments or single vegetative cells that can become airborne. Many of these fungi can grow and reproduce on surfaces within man-made structures and, when disturbed, produce dense biologic aerosols that accumulate within an enclosed space and cause hypersensitivity diseases—such as hypersensitivity pneumonitis, allergic rhinitis, and allergic asthma—and rarely invasive diseases in susceptible people.[30] Many infectious fungal diseases (including coccidioidomycosis, histoplasmosis, blastomycosis, and cryptococcosis) are also known to be transmitted through air carriage of spores or sporebearing soil particles.

Various arthropod particles, including mites and cockroach droppings, have been recovered from air; these are known to be important allergens where they are abundant.[73] Fleas and mosquitoes become airborne through their own actions and can cause discomfort (bites), as well as transmit serious diseases.[139 166 175]

Finally, animal dander and human dander accumulate in any occupied space, and both can be allergenic.

Sources of Biologic Pollutants

Potential sources of biologic aerosols in cabin air include outside air, the cargo compartment, passengers and crew, and structural contamination of the aircraft.

At cruising altitudes, outside air contains very few biologic particles of any kind (a few dark fungal spores per cubic meter of air). These are unlikely to constitute any risk to airline passengers. Outside air that comes in through doorways while a plane is on the ground carries a wide variety of fungal spores, including a few pathogens. In the Southwest, where Coccidioides immitis is endemic, soil particles bearing infective spores often enter enclosures, especially during dust storms. Passengers boarding in these areas would have been exposed in transit to the airport. Passengers and crew stopping over (and not leaving the plane) could receive cabin-associated exposure when doors are open to unload passengers, galley materials, and baggage. Similar situations can arise for other airborne pathogenic fungi. However, control of such exposure would be difficult, because doors must be opened eventually. Most fungi in outdoor air are routinely encountered by everyone and, although allergenic for many, do not constitute a special risk in aircraft cabins. Bacteria are usually not present in outdoor air in concentrations sufficient to cause disease. Exceptions are species of Legionella, soil organisms that multiply in cooling towers and related man-made environments. Infected cooling towers are potential sources of aerosol-borne infection. How far such aerosols can travel in outdoor air and remain infective is unknown.

Cargo compartments can contain animals (which have dander, feces, and urine), arthropods, microorganisms in culture, and contaminated baggage. Aerosols from all these sources could accumulate in a cargo compartment to a point that would be detrimental to human health. However, the potential for contamination of passenger compartments is less clear. Barriers to air circulation between passenger and cargo compartments can range from structural and excellent (Class D) to virtually nonexistent and dependent entirely on airflow patterns (Class B). The greatest danger from cargo sources would be associated with pathogenic microorganisms in cultures that are damaged during transit. The infective dose of some pathogens is a single cell. Pathogens can be transported by mail and are allowed in passenger aircraft if properly packaged. These microorganisms should not be permitted in passenger aircraft, and any nonpathogenic microorganisms in culture should be packed so as to eliminate any possibility of escape.

Primary sources of indoor bacterial and viral aerosols are humans and animals.[141] In addition to bacteria and viruses, clothes-borne fungi and actinomycetes can be carried by people and pets, as can mites. Bacteria are freely shed from human skin with minute skin scales. Clothing contains some of this contamination, and its abrasive action also detaches outer skin layers and increases shedding.[141] Thus, bacteria from this source would be expected to increase during boarding and settling activities and during meal and beverage distribution and to decrease during inactive periods. Occupants can also spread bacteria and viruses by coughing, sneezing, talking, singing, etc. [88] Coughing and sneezing produce the biologically richest aerosols. A sneeze produces very large droplets (200 μm and larger).[141] Immediately on release, respiratory droplets begin to dry. Many become droplet nuclei, which are very small, remain airborne for long periods, and (depending on the organisms and environmental conditions) can remain infective for hours or even days. There is no evidence that the HTLV-III virus, which is associated with acquired immune deficiency syndrome, is transmitted through the air or by casual human contact.

Structural contamination—especially in heating, ventilation, and air-conditioning systems—is of increasing concern.[102] Fungi, actinomycetes, bacteria (including species of Legionella), and protozoans have been found to inhabit such systems in large buildings and have caused widespread disease outbreaks when introduced in aerosol form through the action of the systems themselves. In an aircraft, possible sites of such contamination are the ventilation systems (with their associated filters and water removers) and a wide variety of surfaces, including carpets, upholstered seats, and even metallic surfaces that are persistently or repeatedly wet. Fungi and actinomycetes, in particular, can withstand repeated wet-dry cycles and temperature extremes.

Factors Affecting Airborne Concentrations of Biologic Pollutants and their Health Effects

Factors that can affect airborne concentrations of biologically derived particles include source strength, methods of aerosolization, viability, stability, and ventilation.

Source-strength factors include the number of people and animals in the enclosed space, the number with respiratory or skin infections, and a wide variety of aspects of microbial growth, such as amount of total growth available (which depends on substrate availability, nutrients, water, temperature, and pH), degree of sporulation (which depends on light, temperature, and relative humidity), and spore-cell availability (which depends on viability and colony surface configuration).

Methods of aerosolization include active spore discharge and passive dispersal—coughing, sneezing, talking, air movement, water splashing, and jarring and turbulence. A sneeze produces approximately 2 million viable particles.[141] These do not remain airborne very long, but are highly infective and can be inhaled by people near the infected source. Talking can produce as many as 2,000 particles per explosive sound. Most important for dissemination of fungal aerosols are passive modes, such as air movement over contaminated surfaces. Other biogenic particles—such as skin bacteria, arthropod remains, and dander—are usually introduced into the air through jarring. This can result from human activity (walking on contaminated surfaces, sitting on contaminated seats, or vacuum-cleaning or dusting surfaces) or from jarring of the entire structure (as might be possible in an airplane during turbulent weather). Such antigens as those from fungi, arthropods, and dander do not need to be living to cause hypersensitivity responses, but bacteria and viruses must be viable to be infective.

Among many factors that influence the length of time that bacterial and viral aerosols remain viable are relative humidity, temperature, and time.[120] For viruses, relative humidity and viability are inversely proportional.[91 105] For some bacteria, the situation is reversed: the higher the humidity, the longer the survival. Thus, although the low relative humidities present in most aircraft during flight can be deadly for some bacteria, such conditions probably augment the viability of most viruses. Temperature, which is limiting in extremely cold or extremely hot environments, is unlikely to be a strong factor in an aircraft, where temperature is usually maintained within a comfortable range for the passengers. All organisms die eventually, and each has its own life span. Under ideal conditions, this span can vary among microorganisms, from minutes to many years. Certainly, many microorganisms that cause human disease live long enough even in the stressful environment outside their human hosts to cause disease in enclosed situations.

The ventilation characteristics that directly affect concentrations of biologic particles are the ones that affect concentrations of all interior particles: the quality and quantity of outside air and the quality of filtration in the recirculation systems. The outside air supplied to aircraft cabins during flight is essentially clean. Enough outside air needs to be supplied to dilute the inevitably produced bacterial aerosols to the point where the risk of infection is minimized. Filters currently used in aircraft ventilation systems probably remove only a very small fraction of the continually produced bioaerosols, although data are not available to assess this accurately.

Airborne Concentrations Necessary to Cause Health Effects or Discomfort

Dose-response relations for most organisms are unknown and differ widely from one organism to another. One infectious droplet is sufficient to cause tuberculosis infection, but thousands of droplets are probably necessary to transmit rhinoviruses. In fact, infective dose varies not only with the individual virus of bacterium, but also with such host susceptibility factors as vulnerability of specific cells in the respiratory tract, antibody concentrations, and the presence of predisposing conditions.[79] For example, a person who is in any way immunocompromised—through disease, chemotherapy, or radiation therapy—is highly susceptible to all forms of infection and should not frequent indoor spaces occupied by potentially infectious people. The numbers of spores or particles or concentrations of antigens required to induce hypersensitivity diseases remain completely unknown and most likely vary greatly with the susceptibility of exposed persons.[115]

Available Data

Available Predictive Data from Other Sources

No other environment closely approximates the unusual conditions present in aircraft cabins, but data from a few other sources can be used to predict potential problems aboard aircraft and to provide direction for the design of research. Submarine and spacecraft environments are most nearly like the commercial aircraft environment, except that all their air must be recirculated, because outside air is unavailable. In both, recirculation systems are designed with that in mind, and the quality of air filtration far exceeds that in commercial airliners. In submarines, as many as 30,000 bacteria/ft^3 of air were isolated during sewage handling procedures.[167] However, concentrations generally remain below 20/ft^3, comparable with those in surface ships.[103] Microbiologic measurements were made in the Apollo and Skylab missions.[27] [28] In neither case were concentrations above those expected in other types of interiors. In one Skylab mission, fungal spores were more numerous than expected (but still less numerous than in ground-level outside air), because moldy garments were on board. These data are only marginally relevant to the aircraft environment, because there are major differences in ventilation, air filtration, and passenger load.

Data from doctors' offices and schools clearly indicate that viruses can be circulated through ventilation systems, remain viable, and infect people who have had no physical contact with the source.[22 121] In aircraft cabins, this effect might be augmented by the low relative humidity, which would prolong the life of airborne viruses. It is also apparent from the literature on environmentally tight buildings that microbiologic contamination of heating and ventilation systems can be a serious problem.[120] Although aircraft systems differ substantially from ground-based systems, they have a potential for surface contamination, because temperature differences can cause water to condense and provide suitable substrates for microbial growth.

Available Data on Aircraft

No well-designed research studies that document routine concentrations of microbiologic air contaminants in the aircraft cabin environment have been reported. Studies of other cabin air characteristics either ignore microbiologic contaminants or dismiss them with an unsupported statement that they were "not found." One study, by Air Canada, [37] was carefully designed to assess the risks to healthy passengers and crew associated with transporting passengers with contagious diseases. Bacterial endospores were sprayed from a position five rows from the rear of the aircraft (the position determined, by smoke tracer studies, least likely to cause cabin-wide contamination). During the pretakeoff phase, when the plane was on recirculated air, these spores, although most heavily concentrated near the release site, circulated throughout the cabin and into the cockpit. On takeoff, concentrations away from the source decreased rapidly, but they increased again during descent and landing. The authors stated that background concentrations were low because of the high rate of air-exchange in the cabin, but did not present background data. They concluded that infectious passengers should be carried in the left rear of the aircraft (B-707) and that the engines should be started and forced ventilation begun before such passengers board. Even under those circumstances, it is apparent that a strongly emitting source can contaminate the whole aircraft and that the risk is greatest on the ground and when recirculation is a component of the ventilation.

The risk of contracting epidemic disease on a grounded aircraft is emphasized by a report of an epidemic of influenza directly traceable to a passenger aboard a plane grounded for 4 h in Alaska; 72% of the passengers became ill from the exposure.[104] The absence of other such reports in the literature does not lessen the danger implicit in conditions in a crowded airplane with little or no outside ventilation. This particular epidemic was unusual, in that most patients saw the same physician (in Kodiak), who was in a position to recognize the implications of the situation. If the flight had terminated in Washington, D.C., for example, no physician would have been in a position to recognize even that an epidemic had occurred. Because of the heavy potential risk of spread of infectious disease in aircraft on the ground with no outside ventilation, we recommend that no aircraft with passengers on board remain on the ground without operational forced ventilation for longer than 0.5 h. Open doors are not adequate ventilation sources in this situation. If forced ventilation cannot be initiated within 0.5 h, passengers should be returned to the terminal. This 0.5-h time limit is based on practical consideration of the time required to return a full load of passengers to the terminal. In fact, microbial concentrations will begin to increase as soon as ventilation fails, and risks related to these exposures are unknown.

There are at least four reports of malaria contracted by passengers on aircraft or by visitors to airports shortly after aircraft arrived from areas where malaria was endemic. These cases of malaria very likely resulted from aircraft carriage of malaria-carrying mosquitoes.[139 166 175]

Conclusions and Recommendations

Microbial concentrations have not been measured in aircraft, and therefore accurate risk assessments cannot be made. The Committee feels that microbial aerosols should be measured. A protocol for sampling microbial aerosols in commercial aircraft was developed at the University of Michigan for a pilot study (H. A. Burge, personal communication). This protocol should be tested and expanded into a substantial research effort. Despite the lack of data, the potential for microbial aerosols in aircraft exists, and concentrations can be predicted to be related to ventilation characteristics. Therefore, when ventilation systems are inoperative, passengers should leave the plane within 0.5 h. In addition, physicians should be reminded of the rules stating that infectious passengers must not travel on commercial airliners. It is recognized that many infectious conditions are transmissible long before symptoms appear, rendering this rule relatively valueness, except for severe contagious illnesses. If the risk of infection is to be minimized, the amount of outside air supplied to each passenger during flight should be maximized, because outside air at cruise altitude is essentially clean. The dangers of extensive use of unfiltered, untreated recirculated air should be carefully considered. Every feasible effort should be made to ensure that wet surfaces are prevented or scrupulously cleaned routinely, to prevent structural contamination. Finally, cargo compartments should be kept free of animal excrement, as well as arthropod pests, and pathogenic microorganisms should never be transported on passenger-carrying aircraft.

RELATIVE HUMIDITY

Relative humidity is the ratio of the amount of water vapor in the air at a given temperature to the capacity of the air at that temperature. The term is generally used to mean the percentage of moisture present relative to the amount the air can hold at a given temperature and pressure. The term "vapor pressure" (expressed as millimeters of mercury) refers to the pressure exerted by the (water) vapor on the air mixture at a given temperature and pressure. "Water vapor content" or "water content" (expressed as weight per unit volume or weight of dry air) means the actual amount of water present in the air. All these terms are used in the literature related to relative humidity.

With a constant supply of moisture, relative humidity decreases as temperature increases. The outside air used to ventilate aircraft cabins during flight is very cold and contains very little moisture. On a typical temperate-zone day, ambient air temperature falls from +60°F (+15.6°C to-51.1°C) linearly between sea level and 35,000 ft (10.7 km), and the water content falls from 10 g/kg to less than 0.15 g/kg of dry air. This small amount of moisture plus whatever is accumulated from human sweat, respiration, and cooking activities is all that is available during flight.

Aircraft Ventilation and Relative Humidity

In most aircraft, fresh air is brought in from outside through the engines, cooled, and delivered directly to the cabin with no humidification. Available water from this source, then, remains at about 0.15 g/kg, and, at 20–22°C (68–71.6°F); the relative humidity of the fresh air is less than 1%. The relation between relative humidity and amount of air supplied per passenger is shown in Figure 2–7. Moisture from the passengers themselves will cause relative humidity to increase, depending on the outside-air ventilation rate and the load factor, and it will decrease as rate of outside ventilation increases.

Measured Relative Humidity in Aircraft

Humidity measurements that have been made in aircraft cabins are summarized in Table 5-5. Because of the paucity of data and the failure to indicate outside-air ventilation rates, correlations among relative humidity, load factors, and duration of flight cannot be based on these data. The Lufthansa data[94] (one flight on one aircraft) indicate a fall in relative humidity during flight from 25% to 8.5%. At cabin temperatures of 20–23°C (68–73.4°F), these data suggest that actual water content falls from 4.3 to 1.8 g/m^3 as a function of duration of flight. These values correspond to vapor pressures in the range of 2–6 mm Hg.

TABLE 5-5 Relative Humidity Measured in Aircraft Cabins

		Relative Humidity, %			
Study	Aircraft	Range	Min.	Mean	Max.
Lufthansa[94]	B-747	8.5–25		13	
Applegate[8]	B-747	6.0–40	9.1		16
	DC-10	5.0–34	10.75		22.75
	DC-8	6.0–25	10.5		15.0
	B-727	6.0–16	8.75		12.5
	B-737	17–31	22		25
	Water Content, g/kg of dry air				
Balvanz et al.[19]	B-727		1.5		
	B-737		1.74		
	B-747		1.39		
	B-767		0.739		
	DC-9		2.8		
	DC-10		6.0		

Standards for Other Environments

ASHRAE Standard 55-1981, Thermal Environmental Conditions for Human Occupancy,[5] calls for vapor pressure to range from 5 to 14 mm Hg. The lower end of this scale represents 20% relative humidity at an adjusted dry-bulb temperature of about 72°F (22.2°C).

Effects of Low Relative Humidity on Passengers and Crew

Documented direct effects of low relative humidity on passengers and crew are few. Corneal ulcerations have been reported in wearers of contact lenses after long flights, possibly owing to low oxygen partial pressure, as well as low relative humidity. Hydrophilic contact lenses tend to lose water, but not to the detriment of vision.[46] Removal of contact lenses and the wearing of contact lenses specially designed for use in dry air are successful remedial actions.[40 70] Eng et al.[46] reported that a high percentage of flight attendants complain of dry eyes; however, the value of this questionnaire study is limited because of possible selection bias and lack of controls (see Chapter 6). Of 293 flights on which incidents regarding cabin air quality were reported (presented by the Association of Flight Attendants before a Senate subcommittee), only 27 flights (9%) included complaints of dry eyes, dry throat, dry nose, or dry air. On only three flights were there complaints of "dry air".[15]

Mendez Martin[99] and Kohler[81] showed that urinary calculosis was common among flight personnel, possibly because of low relative humidity, but the Committee found no corroboration of this finding (see Chapter 6).

Reported Health Effects of Low Relative Humidity in Other Environments

Water loss from airways might be an important stimulus of exercise-induced asthma under dry conditions,[7 57] and a slight reduction in lung capacity has been noted in asthmatics at rest in dry environments. In addition, low relative humidity can increase bronchomotor effects of sulfur dioxide in mild asthmatics.[138] At 0% relative humidity, sulfur dioxide at 0.1–0.25 ppm is sufficient to cause a 100% increase in specific airway resistance. This effect has not been studied in the normal population, and synergistic effects between low relative humidity and other pollutants have not been examined. However, low relative humidity itself probably does not cause bronchoconstriction in normal people. In fact, no significant changes in nasal mucus flow rates, nasal or tracheobronchial resistance, or comfort were found in a group of eight healthy men maintained at 9% relative humidity for 79 h.[6] Rappaport et al.[119] indicated that reduced relative humidity (15–40%) might be beneficial to pollen-asthma sufferers. Abrupt changes in relative humidity (more than 10%) can cause discomfort in some people, but re-equilibration tends to occur within about 15 min.

Evidence on the common belief that low relative humidity increases the risk of respiratory infection is conflicting. The Baetjer studies [17] with mice and chicks found some correlation between virus titers and low humidity with high temperature, a combination not common in aircraft. Mucociliary clearance rates in chicks exposed to a range of temperatures and vapor pressures were increased at low vapor pressures; most of these experiments failed to duplicate aircraft cabin temperatures and humidity. Baetjer commented on possible effects of temperature and vapor pressure on the skin (including absorption of noxious chemicals), but presented no data to indicate that low vapor pressures diminish or amplify skin absorption.

Melia et al.[98] reported a positive correlation of respiratory infections with relative humidity, i.e., as humidity increased, respiratory infections increased; but the relative humidity was higher than that in aircraft. The one disease state in which high relative humidity has been thought to be at least palliative, croup, has been shown to be unaffected by even very high humidities.[23] It has been shown that people overwintering at the Antarctic station, where indoor relative humidities approach those in aircraft cabins, are not at increased risk of respiratory infections after their return to a more temperate environment.[72] These data are only marginally relevant to cabin air quality, because there are major differences in conditions and patterns of exposure. Lidwell et al.[89] demonstrated that increased incidence of respiratory infection in winter is related to temperature, not to relative humidity, and stated that this analysis "contradicts any arguments based on virus survival in relation to indoor humidity or on a postulated damaging effect, due to drying, on the mucous membranes, predisposing to the initiation of infection when the indoor humidity falls in cold weather." However, after a less extensive study, Gelperin[54] reported that the rate of upper respiratory infection was 5–10% lower in humid barracks (relative humidity, 40%) than in unhumidified barracks (relative humidity, 20%). He reported an equivalent decrease in foot infections that is difficult to explain. Green et al.[56] reported data from Saskatoon schools that indicated slightly lower absenteeism (4.6% vs. 5.1%) in schools with slightly higher relative humidity (overall relative humidity range, 18.4–38.6%); the data show a marginal effect of relative humidity in a range not comparable with that found in aircraft cabins.

Low relative humidity has been shown to be a factor in dry, scaly skin rashes when specific irritants are also present[173] and to contribute to winter drying and chapping of hands and face when accompanied by exposure to excess cold or water. Assuming that flight attendants wash their hands often, dry aircraft air could cause increased problems with rough dry hands, but would be unlikely to affect passengers, who are exposed relatively briefly.

Indirect effects of relative humidity on passengers and crew are associated primarily with the viability of microorganisms. Humidification of schools has been considered important for many years as a means of decreasing virus survival, but decreased concentrations of viruses in humid school air have not been documented. It has been shown that concentrations of some bacteria can be directly related to relative humidity: as humidity decreases, bacterial concentrations decrease. As mentioned earlier, indoor fungal spore concentrations are also directly related to relative humidity.

Summary

The health risks associated with clean, dry air appear quite low, especially for normal people, and probably do not justify the cost and potential microbiologic complications that would attend installation of active humidification systems in aircraft.

PRESSURIZATION

The Committee recognizes that pressurization of the cabin to an equivalent altitude of 5,000–8,000 ft is physiologically safe—no supplemental oxygen is needed to maintain sufficient arterial oxygen saturation. The percentage of oxygen remains virtually unchanged (21%) at all altitudes. But partial pressures of gases change, as shown in Table 5–6, where the partial pressure of oxygen is shown to decrease from 160 mm Hg at sea level to 110 mm Hg at 10,000 ft. The oxygen-hemoglobin dissociation figures, which indicate the amount of oxygen held by hemoglobin at various partial pressures of oxygen, clearly show that the dissociation of oxygen from hemoglobin decreases with decreasing partial pressure of oxygen. There is a decrement in night vision at 4,000–6,000 ft, and a 7–10% decrement in maximal performance at altitudes between 7,000 and 10,000 ft. Cabin altitudes can legally reach 8,000 ft, but after failure of the pressurization system could reach as high as 15,000 ft. At these altitudes, people with advanced cardiopulmonary disease might be at some risk.

TABLE 5-6 Hypobaric Pressure and Arterial Oxygen Saturationa

Pressure Altitude, ft	Atmospheric Pressure, mm Hg	Oxygen Partial Pressure,[b] mm Hg	Arterial Oxygen Saturation Without Supplemental Oxygen, %
0	760	160	96
2,500	694	147	95
5,000	632	133	95
7,500	575	121	93
10,000	523	110	89

[a] Data from Mohler.[101]

[b] 20% of atmospheric pressure.

The normal rates of change of cabin pressure (500 ft/min in increasing altitude and 300 ft/min in decreasing altitude) do not pose a problem for passengers in normal health. However, persons suffering from upper respiratory infections might experience pain of varied severity, temporary loss of hearing, and tinnitus due to inflammation of the nasopharyngeal orifice of the eustachian tube or due to swelling of mucous membranes in the sinus ostia. Cabin crew members should be advised as to the symptoms and procedures to alleviate them.

The Committee sought evidence of operating practices in which an aircraft was pressurized at altitudes above 8,000 ft, but could not find any records that would confirm the existence of such a practice. Because operation of the pressure control system on modern jet aircraft is usually fully automatic, the likelihood of excursions above 8,000 ft is small. In addition, regulations require that an audible or visible warning be given to the crew if the cabin altitude exceeds 10,000 ft. Nevertheless, the Committee believes that systematic measurement of cabin pressure on a representative sample of routine commercial flights would be advisable. The information gathered could be used to assess the adequacy of current requirements or establish a basis for regular monitoring, if necessary.

The Committee recognizes that properly informing the public about health risks is complicated and difficult. Many physicians advise their patients about health risks of flying, but passengers do not always consult physicians about their travel plans, so this method is not as effective as it should be. Carefully worded messages could be made available to potential airline passengers with acute or chronic middle ear problems, heart problems, or lung disease. This could be accomplished through provision of the information at ticket counters or through travel agents and others who sell tickets.

CARBON DIOXIDE

Carbon dioxide is the product of normal human metabolism, which is the predominant source in aircraft cabins. The carbon dioxide concentration in the cabin depends on the ventilation rate, the number of people present, and their individual rates of carbon dioxide production, which vary with activity and (to a smaller degree) with diet and health.

Federal Aviation Regulations[163] specify that "carbon dioxide in excess of 3 percent by volume (sea level equivalent) is considered hazardous in the case of crewmembers. Higher concentrations of carbon dioxide may be allowed in crew compartments if appropriate protective breathing equipment is available" (see Chapter 3). In contrast, ASHRAE Standard 62-1981, <u>Ventilation for Acceptable Air Quality</u>,[4] bases indoor ventilation requirements on a carbon dioxide production rate of 0.63 cubic foot per hour (cfh) per person and outside air containing 0.03% carbon dioxide. Ventilation calculations are based on a limit of 0.5% carbon dioxide. However, as an additional safety factor to cover individual activity levels and diet and health variations, a recommended limit of 0.25% carbon dioxide is used, and that establishes a ventilation rate of 5 cfm/person. In comparison, the environmental exposure limit adopted in 1984–1985 by the American Conference of Governmental Industrial Hygienists (ACGIH) gives 5,000 ppm as the time-weighted average (TWA).[2] The TWA is the concentration, for a normal 8-h workday and a 40-h workweek, to which nearly all workers can be repeatedly exposed, day after day, without adverse effects. The Occupational Safety and Health Administration also lists the 5,000-ppm concentration as the TWA for workers.[149] ACGIH has a short-term exposure limit (STEL) of 15,000 ppm, although it has issued a notice of intended change to 30,000 ppm. A STEL is defined as a 15-min TWA exposure that should not be exceeded at any time during a working day.

Under normal conditions, carbon dioxide at 0.6% has little effect on lung function, increasing it only about 10% above normal. As carbon dioxide concentration increases, there is an increase in both the rate and the depth of breathing, which reaches twice normal at 3% carbon dioxide. At that concentration, there is some discomfort; as it increases, headache, malaise, and fatigue occur, and the air is reported as stale. At altitudes up to 8,000 ft, the reduced pressure has no significant effect on the symptoms or other response to increased carbon dioxide concentrations.

In discussing federal regulations, Chapter 3 pointed out that the current FAR concerning acceptable cabin concentrations of carbon dioxide is several decades old. The Committee finds that there is a need to consider revision of this standard and recommends that FAA review it in the light of more recent scientific findings and in comparison with standards established for air quality in buildings occupied by the general public and with workplace exposure limits adopted by ACGIH. If any potential for hazardous concentrations is discovered, the analysis should be supported by appropriate testing.

OTHER POTENTIAL EXPOSURES

Aircraft occupants can be exposed to a number of pollutants from materials used to construct or maintain the cabin. These include volatile organic chemicals emitted by materials used in furnishing the cabin, pesticides, and cleaning agents.

Volatile organic chemicals in the aircraft cabin have numerous potential sources: adhesives, lubricants, elastomers, sealing compounds, coatings, etc., used in the construction or maintenance of the cabin interior. Some of these products have been tested for offgassing of volatile chemicals as part of a government space study.[117] The offgassing chemicals include acetone, ethanol, benzene, toluene, and n-butanol. Many of them have serious toxicity (for example, benzene is a known human carcinogen). However, exposure to them would be affected by the type and amount of the offgassing products used in the aircraft, the rate of offgassing under the conditions of use, the age of the products, and the ventilation rate in the aircraft. The Committee could find no monitoring data on the concentrations of volatile organic chemicals in aircraft cabins during operation.

Insecticides can be used on aircraft to control pests of public-health or agricultural importance. The Centers for Disease Control (CDC) Division of Quarantine is authorized to require removal of insects from aircraft leaving foreign areas that are infected with insect-borne communicable disease, if the aircraft are suspected of harboring insects of public-health importance.[43] Insecticides used by the airlines must be approved by CDC. In 1979, it approved resmethrin (2% aerosol) and d-phenothrin (2% aerosol) for use in aircraft.[152] However, CDC does not now require the use of pesticides on aircraft (B. Coull, personal communication, 1986). If an aircraft is suspected of harboring insects of particular agricultural importance (notably the Japanese beetle), the Department of Agriculture can require fumigation.[71] d-Phenothrin is used in some aircraft traveling from regulated airports in the eastern United States to protected areas in the western United States, when inspectors deem it advisable.

d-Phenothrin and resmethrin, like other pyrethroids, are neither skin irritants nor skin sensitizers. Inhalation toxicity and dermal toxicity are fairly low. Neither is teratogenic in rats, mice, or rabbits or mutagenic in various bacterial strains.[100] Both are highly effective in killing some species of flies, mosquitoes, and other flying insects.[34 132] In a test of physical and insecticidal properties of (+)-phenothrin, Liljedahl et al.[90] found that none of the cabin crew or scientists who were present experienced any odor or irritation during the use of the insecticide or from the residual deposit. Sullivan et al.,[147] in worldwide aircraft trials, found resmethrin to be acceptable to 92% of the passengers questioned after the cabin was sprayed.

Although no health problems among cabin occupants or crew have been reported as a result of the use of insecticides, there is always a potential for health problems due to their improper application.

Disinfectants and cleaning chemicals used on aircraft can also be a source of exposure to occupants and might cause health problems in some people. For example, carpet shampoos have been linked to outbreaks of respiratory illness among people exposed soon after shampoo application.[83 122] Although the Committee has no data or other documentation on health problems caused by these types of chemicals used on aircraft, it is reasonable to assume that they could occur on aircraft if appropriate precautions were not followed.

REFERENCES

1. Adlkofer, F., G. Scherer, and H. Weimann. Small-airways dysfunction in passive smokers. N. Engl. J. Med. 303:392, 1980. (letter)
2. American Conference of Governmental Industrial Hygienists. Carbon dioxide, pp. 69–70. In Documentation of the Threshold Limit Values. 4th ed. Cincinnati, Ohio: American Conference of Governmental Industrial Hygienists, 1984.
3. American Lung Association. Americans Want Smoke-Free Air at Work. New York, N.Y.: American Lung Association, Dec. 5, 1985.

4. American Society of Heating, Refrigerating, and Air-Conditioning Engineers, Inc. ASHRAE Standard: Ventilation for Acceptable Air Quality. ASHRAE 62-1981. Atlanta, Ga.: American Society of Heating, Refrigerating, and Air-Conditioning Engineers, Inc., 1981.
5. American Society of Heating, Refrigerating, and Air-Conditioning Engineers, Inc. Thermal Environmental Conditions for Human Occupancy. ANSI/ASHRAE 55-1981. Atlanta, Ga.: American Society of Heating, Refrigerating and Air-Conditioning Engineers, Inc., 1981.
6. Andersen, I., G. R. Lundqvist, P. L. Jensen, and D. F. Proctor. Human response to 78-hour exposure to dry air. Arch. Environ. Health 29:319–324, 1974.
7. Anderson, S. D., R. E. Schoeffel, R. Follet, C. P. Perry, E. Daviskas, and M. Kendall. Sensitivity to heat and water loss at rest and during exercise in asthmatic patients. Eur. J. Respir. Dis. 63:459–471, 1982.
8. Applegate, W. C. Aircraft Cabin Air Quality Study. (unpublished document, United Airlines, 1982)
9. Araki, Y., and K. Kasano. Building control law and dust collectors. Mitsubishi Denki Giho 48:470–477, 1974. (in Japanese)
10. Aronow, W. S. Aggravation of angina pectoris by two percent carboxyhemoglobin. Am. Heart J. 101:154–157, 1981.
11. Aronow, W. S. Effect of carbon monoxide on cardiovascular disease. Prev. Med. 8:271–278, 1979.
12. Aronow, W. S. Effect of passive smoking on angina pectoris. N. Engl. J. Med. 299:21–24, 1978.

13. Aronow, W. S., J. Cassidy, J. S. Vangrow, H. March, J. C. Kern, J. R. Goldsmith, M. Khemka, J. Pagano, and M. Vawter. Effect of cigarette smoking and breathing carbon monoxide on cardiovascular hemodynamics in anginal patients. Circulation 50:340–347, 1974.
14. Aronow, W. S., and M. W. Isbell. Carbon monoxide effect on exercise-induced angina pectoris. Ann. Intern. Med. 79:392–395, 1973.
15. Association of Flight Attendants. Exhibit A (passenger comments and flight attendant symptoms, 01/77–01/82). In U.S. Senate, Committee on Commerce, Science, and Transportation, Subcommittee on Aviation (97th Congress, 2nd Session). Airliner Cabin Safety and Health Standards: Hearing on S. 1770, May 20, 1982. Washington, D.C.: U.S. Government Printing Office, 1982.
16. Aviado, D. M. Small-airways dysfunction in passive smokers. N. Engl. J. Med. 303:393, 1980. (letter)
17. Baetjer, A. M. Role of environmental temperature and humidity in susceptibility to disease. Arch. Environ. Health 16:565–570, 1968.
18. Baily, P. E. Neutron radiation dosimetry in high altitude flight personnel. Aviat. Space Environ. Med. 53:808–817, 1982.
19. Balvanz, J., S. C. Bowman, A. Fobelets, T. Lee, K. Papamichael, and R. Yoder. Examination of the Cabin Environment of Commercial Aircraft. Ames, Iowa: Iowa State University, 1982.
20. Banaszak, E. F., J. Barboriak, J. Fink, G. Scanlon, D. Schlueter, A. Sosman, W. Thiede, and G. Unger. Epidemiologic studies relating thermophilic fungi and hypersensitivity lung syndromes. Am. Rev. Respir. Dis. 110:585–591, 1974.

21. Barth, D. S., J. C. Romanovsky, J. H. Knelson, A. P. Altshuller, and R. J. Horton. National air quality standards for automative pollutants: A critical review. J. Air Pollut. Control Assoc. 21:544–548, 1971.
22. Bloch, A. B., W. A. Orenstein, W. M. Ewing, W. H. Spain, G. F. Mallison, K. L. Herrmann, and A. R. Hinman. Measles outbreak in a pediatric practice: Airborne transmission in an office setting. Pediatrics 75:676–683, 1985.
23. Bourchier, D., K. P. Dawson, and D. M. Fergusson. Humidification in viral croup: A controlled trial. Aust. Paediatr. J. 20:289–291, 1984.
24. Brabets, R. I., C. K. Hersh, and M. J. Klein. Ozone measurement survey in commercial jet aircraft. J. Aircraft 4:59–64, 1967.
25. Bramlitt, E. T. Commercial aviation crew member radiation doses. Health Phys. 49:945–948, 1985.
26. Bramlitt, E. T. Radiation safety requirements for crewmembers: Petition for rulemaking. Federal Register 49:31298, 1984.
27. Brockett, R. M., and J. K. Ferguson. Microbiological sampling of the spacecraft atmosphere during a simulated Skylab mission. Aviat. Space Environ. Med. 46:30–32, 1975.
28. Brockett, R. M., J. K. Ferguson, and M. R. Henney. Prevalence of fungi during Skylab missions. Appl. Environ. Microbiol. 36:243–246, 1978.
29. Brunekreef, B., P. Fischer, B. Remijn, R. van der Lende, J. Schouten, and P. Quanjer. Indoor air pollution and its effect on pulmonary function of adult non-smoking women. III. Passive smoking and pulmonary function. Int. J. Epidemiol. 14:227–230, 1985.

30. Burge, H. A. Indoor sources for airborne microbes in indoor air and human health, pp. 139–148. In R. B. Gammage and S. V. Kaye, Eds. Indoor Air and Human Health. Chelsea, Mich.: Lewis Publishers, 1985.
31. Cabin ozone concentrations. Code of Federal Regulations, Title 14, Pt. 25.832. Washington, D.C.: U.S. Government Printing Office, 1985.
32. Cain, W. S., B. P. Leaderer, R. Isseroff, L. G. Berglund, R. J. Huey, E. D. Lipsitt, and D. Perlman. Ventilation requirements in buildings. I. Control of occupancy odor and tobacco smoke odor. Atmos. Environ. 17:1183–1197, 1983.
33. Cain, W. S., T. Tosun, L. C. See, and B. Leaderer. Environmental tobacco smoke: Sensory reactions of occupants. Atmos. Environ. 1986. (in press)
34. Cawley, B. M., W. N. Sullivan, M. S. Schechter, and J. U. McGuire. Desirability of three synthetic pyrethroid aerosols for aircraft disinsection. Bull. WHO 51:537–540, 1974.
35. Chan, W. C., and S. C. Fung. Lung cancer in non-smokers in Hong Kong, pp. 199–202. In E. Grundmann, J. Clemmesen, and C. S. Muir, Eds. Geographical Pathology in Cancer Epidemiology. Cancer Campaign. Vol. 6. Cancer Epidemiology. Stuttgart: Gustaav Fischer Verlag, 1982.
36. Clausen, G. H., P. O. Fanger, W. S. Cain, and B. P. Leaderer. Stability of tobacco smoke odor in enclosed spaces, pp. 437–441. In B. Berglund, T. Lindvall, and J. Sundell, Eds. Indoor Air: Sensory and Hyperreactivity Reactions to Sick Buildings. Vol. 3. Stockholm: Swedish Council for Building Research, 1984.
37. Clayton, A. J., D. C. O'Connell, R. A. Gaunt, and R. E. Clarke. Study of the microbiological environment within long-and medium-range Canadian Forces aircraft. Aviat. Space Environ. Med. 47:471–472, 1976.

38. Cohen, D., S. F. Arai, and J. D. Brain. Smoking impairs long-term dust clearance from the lung. Science 204:514–517, 1979.
39. Comstock, G. W., M. B. Meyer, K. J. Helsing, and M. S. Tockman. Respiratory effects on household exposures to tobacco smoke and gas cooking. Am. Rev. Respir. Dis. 124:143–148, 1981.
40. Corboy, J. M., and J. C. Tannehill. Effects of air travel on contact lens wearers . Am. J. Ophthalmol. 76:166–167, 1973.
41. Correa, P., L. W. Pickle, E. Fontham, Y. Lin, and W. Haenszel. Passive smoking and lung cancer. Lancet 2:595–597, 1983.
42. Dahms, T. E., J. F. Bolin, and R. G. Slavin. Passive smoking: Effects on bronchial asthma. Chest 80:530–534, 1981.
43. Disinsection of aircraft. Code of Federal Regulations, Title 42, Pt. 71.44. Washington, D.C.: U.S. Government Printing Office, 1985.
44. Duncan, D. B., and P. P. Greaney. Passive smoking and uptake of carbon monoxide in flight attendants. JAMA 251:2658–2659, 1984. (letter)
45. Elliott, L. P., and D. R. Rowe. Air quality during public gatherings. J. Air Pollut. Control Assoc. 25:635–636, 1975.
46. Eng, W. G., L. K. Harada, and L. S. Jagerman. The wearing of hydropholic contact lenses aboard a commercial jet aircraft. I. Humidity effects on fit. Aviat. Space Environ. Med. 53:235–238, 1982.
47. Ferin J., G. Urbankova, and A. Vlckova. Influence of tobacco smoke on the elimination of particles from the lungs. Nature 206:515–516, 1965.
48. Flight time limitations: All flight crewmembers. Code of Federal Regulations, Title 14, Pt. 121.471. Washington, D.C.: U.S. Government Printing Office, 1985.

49. Foliart, D., N. L. Benowitz, and C. E. Becker. Passive absorption of nicotine in airline flight attendants. New Engl. J. Med. 308:1105, 1983. (letter)
50. Folinsbee, L. J., B. L. Drinkwater, J. F. Bedi, and S. M. Horvath. The influence of exercise on the pulmonary function changes due to exposure to low concentrations of ozone, pp. 125–145. In L. J. Folinsbee, Ed. Environmental Stress: Individual Human Adaptations. New York, N.Y.: Academic Press, 1978.
51. Gallup Organization, Inc. 1985 Air Travel Survey. Washington, D.C.: Air Transport Association of America, 1985.
52. Garfinkel, L., O. Auerbach, and L. Joubert. Involuntary smoking and lung cancer: A case-control study. JNCI 75:463–469, 1985.
53. Garland, C., E. Barrett-Connor, L. Suarez, M. H. Criqui, and D. L. Wingard. Effects of passive smoking on ischemic heart disease mortality of nonsmokers. Am. J. Epidemiol. 121:645–650, 1985.
54. Gelperin, A. Humidification and upper respiratory infection incidence. Heating/Piping/Air-Conditioning 45(3):77–78, 1973.
55. Gillis, G. R., D. J. Hole, V. M. Hawthorne, and P. Boyle. The effect of environmental tobacco smoke in two urban communities in the west of Scotland. Eur. J. Respir. Dis. 65 (Suppl. 133):121–126, 1984.
56. Green, G. H., K. Sareen, and A. D. Osborne. The effect of ventilation and relative humidity upon airborne bacteria in schools. ASHRAE Trans. 91(2a):13–28, 1985.
57. Hahn, A., S. D. Anderson, A. R. Morton, J. L. Black, and K. D. Fitch. A reinterpretation of the effect of temperature and water content of the inspired air in exercise-induced asthma. Am. Rev. Respir. Dis. 130:575–579, 1984.

58. Halfpenny, P. F., and P. S. Starrett. Control of odor and irritation due to cigarette smoking aboard aircraft. ASHRAE J. 3(3):39–44, 1961.
59. Heuss, J. M., G. J. Nebel, and J. M. Colucci. National air quality standards for automotive pollutants: A critical review. J. Air Pollut. Control Assoc. 21:535–544, 1971.
60. Hiller, F. C. Deposition of sidestream cigarette smoke in the human respiratory tract. Prev. Med. 13:602–607, 1984.
61. Hiller, F. C., K. T. McCusker, M. K. Mazumder, J. D. Wilson, and R. C. Bone. Deposition of sidestream cigarette smoke in the human respiratory tract. Am. Rev. Respir. Dis. 124:406–408, 1982.
62. Hirayama, T. Cancer mortality in nonsmoking women with smoking husbands based on a large-scale cohort study in Japan. Prev. Med. 13:680–690, 1984.
63. Hirayama, T. Non-smoking wives of heavy smokers have a higher risk of lung cancer: A study from Japan. Br. Med. J. 282:183–185, 1981.
64. Hoffmann, D., N. J. Haley, K. D. Brunnemann, J. D. Adams, and E. L. Wynder. Cigarette Sidestream Smoke: Formation, Analysis and Model Studies on the Uptake by Nonsmokers. U.S. Japan Meeting, New Etiology of Lung Cancer, Honolulu, Hawaii, March 21–23, 1983.
65. Horvath, S. M., J. A. Gliner, and L. J. Folinsbee. Adaptation to ozone: Duration of effect. Am. Rev. Respir. Dis. 123:496–499, 1981.
66. Horvath, S. M., J. A. Gliner, and J. A. Matsen-Twisdale. Pulmonary function and maximum exercise responses following acute ozone exposure. Aviat. Space Environ. Med. 50:901–905, 1979.

67. Huber, G. L. Small-airways dysfunction in passive smokers. N. Engl. J. Med. 303:392–393, 1980. (letter)
68. International Commission on Radiological Protection. Statement from the 1983 Meeting of the International Commission on Radiological Protection. Ann. ICRP 14(1):v–vii, 1984.
69. International Radiological Protection Commission, Task Group on Lung Dynamics. Deposition and retention models for internal dosimetry of the human respiratory tract. Health Physics 12:173–207, 1966.
70. Jagerman, L. S. Effects of air travel on contact-lens wearers. Am. J. Ophthalmol. 75:533, 1973.
71. Japanese beetle. Code of Federal Regulations, Title 7, Pt. 301.48. Washington, D.C.: U.S. Government Printing Office, 1985.
72. Jennings, L. C., and J. L. Faoagali. A study of the susceptibility of Antarctic winter-over personnel to viral respiratory illness on their return to New Zealand. N. Z. Antarctic Record 3(1):29–30, 1980.
73. Kang, B., and J. L. Chang. Allergenic impact of inhaled arthropod material. Clin. Rev. Allergy 3:363–375, 1985.
74. Kauffmann, F., D. W. Dockery, F. E. Speizer, and B. G. Ferris, Jr. Respiratory symptoms and lung function in women with passive and active smoking. Am. Rev. Respir. Dis. 133:A157, 1986.
75. Kauffman, F., J. F. Tessier, and P. Oriol. Adult passive smoking in the home environment: A risk factor for chronic airflow limitation. Am. J. Epidemiol. 117:269–280, 1983.
76. Kentner, M., G. Triebig, and D. Weltle. The influence of passive smoking on pulmonary function: A study of 1,351 office workers. Prev. Med. 13:656–669, 1984.

77. Kerka, W. F., and C. M. Humphreys. Temperature and humidity effect on odor perception. ASHRAE Trans. 62: 531–552, 1956.
78. Klus, H., and H. Kuhn. Distribution of various tobacco-smoke components among mainstream and sidestream smoke: A survey. Beitr. Tabakforsch. 11:229–265, 1982.
79. Knight, V. Airborne transmission and pulmonary deposition of respiratory viruses, pp. 1–9. In V. Knight, Ed. Viral and Mycoplasmal Infections of the Respiratory Tract. Philadelphia, Pa.: Lea & Febiger, 1973.
80. Knoth, A., H. Bohn, and F. Schmidt. Passive smoking as cause of lung cancer in women non-smokers. Med. Klin. 78(2):66–69, 1983. (in German)
81. Kohler, F. P. Asymptomatic upper urinary tract lithiasis in flying personnel. J. Urol. 86:370–372, 1961.
82. Koo, L. C., J. H-C. Ho, and D. Saw. Is passive smoking an added risk factor for lung cancer in Chinese women? J. Exp. Clin. Cancer Res. 3:277–283, 1984.
83. Kreiss, K., M. G. Gonzales, K. L. Conright, and A. R. Scheere. Respiratory irritation due to carpet shampoo: Two outbreaks. Environ. Int. 8:337–341, 1982.
84. Kulle, T. J., L. R. Sauder, J. R. Hebel, and M. D. Chatham. Ozone response relationships in healthy nonsmokers. Am. Rev. Respir. Dis. 132:36–41, 1985.
85. Lategola, M. T., C. E. Melton, and E. A. Higgins. Effects of ozone on symptoms and cardiopulmonary function in a flight attendant surrogate population. Aviat. Space Environ. Med. 51:237–246, 1980.

86. Lategola, M. T., C. E. Melton, and E. A. Higgins. Pulmonary and symptom threshold effects of ozone in airline passengers and cockpit crew surrogates. Aviat. Space Environ. Med. 51:878–884, 1980.
87. Lee, P. N. Passive smoking. Lancet 1:791, 1982. (letter)
88. Letts, R. M., and E. Doermer. Conversation in the operating theater as a cause of airborne bacterial contamination. J. Bone Joint Surg. 65:357–362, 1983.
89. Lidwell, O. M., R. W. Morgan, and R. E. O. Williams. The epidemiology of the common cold. IV. The effect of weather. J. Hyg. Camb. 63:427–439, 1965.
90. Liljedahl, L. A., H. J. Retzer, W. N. Sullivan, M. S. Schechter, B. M. Cawley, N. O. Morgan, C. M. Amyx, B. A. Schiefer, E. J. Gerberg, and R. Pal. Aircraft disinsection: The physical and insecticidal characteristics of (+)-phenothrin applied by aerosol at "blocks away." Bull. WHO 54:391–396, 1976.
91. Loosli, C. G., H. M. Lemon, O. H. Robertson, and E. Appel. Experimental air-borne influenza infection. I. Influence of humidity on survival of virus in air. Proc. Soc. Exp. Biol. Med. 53:205–206, 1943.
92. Los Angeles County Air Pollution Control District. Study of Jet Aircraft Emissions and Air Quality in the Vicinity of the Los Angeles International Airport. APTD-0662. Research Triangle Park, N.C.: U.S. Environmental Protection Agency, Air Pollution Control Office, 1971. (available from NTIS as PB-198699)
93. Lozano, E. R., W. W. Melvin, Jr., and S. Hochheiser. Air pollution emissions from jet engines. J. Air Pollut. Control Assoc. 18:392–394, 1968.

94. Lufthansa German Airlines. Report on Air Quality Study with Half Pack Operation: 747. (unpublished, 1982)
95. MacMahon, B. Prenatal x-ray exposure and childhood cancer. JNCI 28:1173–1191, 1962.
96. MacMahon, B., and G. B. Hutchison. Prenatal x-ray and childhood cancer: A review. Acta Un. Int. Cancr. 20:1172–1174, 1964.
97. McDonnell, W. F., D. H. Horstman, M. J. Hazucha, E. Seal, Jr., E. D. Haak, S. A. Salaam, and D. E. House. Pulmonary effects of ozone exposure during exercise: Dose-response characteristics. J. Appl. Physiol. 54:1345–1352, 1983.
98. Melia, R. J. W., C. du V. Florey, R. W. Morris, B. D. Goldstein, H. H. John, D. Clark, I. B. Craighead, and J. C. Mackinlay. Childhood respiratory illness and the home environment. II. Association between respiratory illness and nitrogen dioxide, temperature and relative humidity. Int. J. Epidemiol. 11:164–169, 1982.
99. Mendez Martin, A. Urolithiasis among flight personnel. Med. Clin. (Barcelona):72:65–70, 1979. (in Spanish)
100. Miyamoto, J. Degradation, metabolism and toxicity of synthetic pyrethroids. Environ. Health Perspect. 14:15–28, 1976.
101. Mohler, S. R. Physiologically Tolerable Decompression Profiles for Supersonic Transport Type Certification. AM 70-12. Washington, D.C.: U.S. Federal Aviation Administration, Office of Aviation Medicine, 1970.
102. Morey, P. R., M. J. Hodgson, W. G. Sorenson, G. J. Kullman, W. W. Rhodes, and G. S. Visvesvara. Environmental studies in moldy office buildings: Biological agents, sources, and preventive measures . Ann. Am. Conf. Gov. Ind. Hyg. 10:21–35, 1984.

103. Morris, J. E. W. Microbiology of the submarine environment. Proc. R. Soc. Med. 65:799–800, 1972.
104. Moser, M. R., T. R. Bender, H. S. Margolis, G. R. Noble, A. P. Kendal, and D. G. Ritter. An outbreak of influenza aboard a commercial airliner. Am. J. Epidemiol. 110:1–6, 1979.
105. Muchmore, H. G., A. J. Parkison, J. E. Humphries, E. N. Scott, D. A. McIntosh, L. V. Scott, M. K. Cooney, and J. A. R. Miles. Persistent parainfluenza virus shedding during isolation at the South Pole. Nature 289:187–189, 1981.
106. Murray, A. L., and P. S. Lawrence. Sequelae to smoking cessation: A review. Clin. Psychol. Rev. 4:143–157, 1984.
107. Nastrom, G. D., J. D. Holdeman, and P. J. Perkins. Measurements of cabin and ambient ozone on B747 airplanes. J. Aircraft 17: 246–249, 1980.
108. National Council on Radiation Protection and Measurements. A Review of NCRP Radiation Dose Limit for Embryo and Fetus in Occupationally-Exposed Women. NCRP Report 53. Washington, D.C.: National Council on Radiation Protection and Measurements, 1977.
109. National Council on Radiation Protection and Measurements. Natural Background Radiation in the United States. NCRP Report 45. Washington, D.C.: National Council on Radiation Protection and Measurements, 1975.
110. National Research Council, Committee on Indoor Pollutants. Indoor Pollutants. Washington, D.C.: National Academy Press, 1981.
111. Newman, W. H., and M. R. Viele. Engine Bleed Air Reduction in DC-10. NASA-CR-159846. Long Beach, Cal.: Douglas Aircraft Company, 1980.

112. Oakley, D. T. Natural Radiation Exposure in the United States. ORP/SID 72-1. Washington, D.C.: U.S. Environmental Protection Agency, Office of Radiation Programs, 1972.
113. Ostberg, O., and R. Mills-Orring. Cabin Attendants' Working Environment: A Questionnaire Study. Technical Report 180:74 T. Lulea: University of Lulea, Department of Human Work Sciences, 1980. (in Swedish)
114. Perkins, P. J., J. D. Holdeman, and D. J. Gauntner. Aircraft cabin ozone measurements on B747-100 and B747-SP aircraft: Correlations with atmospheric ozone and ozone encounter statistics. NASA TM-79060. Cleveland, Ohio: NASA Lewis Research Center, 1978.
115. Platts-Mills, T. A., F. Rawle, and M. D. Chapman. Problems in allergen standardization. Clin. Rev. Allergy 3:271–290, 1985.
116. Pregnancy Discrimination Act of 1978: An Act to Amend Title VII of the Civil Rights Act of 1964 to Prohibit Sex Discrimination on the Basis of Pregnancy. PL 95-555. U.S. Statutes at Large 92:2076–2077, 1978.
117. Pustinger, J. V., Jr., and F. N. Hodgson. Identification of Volatile Contaminants of Space Cabin Materials. AMRL-TR-67-58. Dayton, Ohio: Monsanto Research Corporation, 1967. (available from NTIS as N67-37551)
118. Quigley, S. Jet engine emissions. J. Ir. Med. Assoc. 65:258–262, 1972.
119. Rappaport, B. Z., T. Nelson, and W. H. Welker. The effect of low relative humidity at constant temperature on pollen asthma. J. Allergy 6:111–122, 1935.
120. Riley, R. L. Indoor airborne infection. Environ. Int. 8:317–320, 1982.

121. Riley, R. L. Indoor spread of respiratory infection by recirculation of air. Bull. Physiopathol. Respir. 15:699–705, 1979.
122. Robinson, P. A., R. V. Tauxe, W. G. Winkler, and M. E. Levy. Respiratory illness in conference participants following exposure to rug shampoo. Infection Control 4:158–160, 1983.
123. Rogers, J. W. Results of FAA Cabin Ozone Monitoring Program in Commercial Aircraft in 1978 and 1979. FAA-EE-80-10. Washington, D.C.: U. S. Federal Aviation Administration, Office of Environment and Energy, 1980.
124. Sakuma, H., M. Kusama, S. Munakata, T. Ohsumi, and S. Sugawara. The distribution of cigarette-smoke components between mainstream and sidestream smoke. 1. Acidic components. Beitr. Tabakforsch. 12:63–71, 1983.
125. Sakuma, H., M. Kusama, K. Yamaguchi, T. Matsuki, and S. Sugawara. The distribution of cigarette-smoke components between mainstream and sidestream smoke. 2. Bases. Beitr. Tabakforsch. 12:199–209, 1984.
126. Sakuma, H., M. Kusama, K. Yamaguchi, and S. Sugawara. The distribution of cigarette-smoke components between mainstream and sidestream smoke. 3. Middle and higher boiling components. Beitr. Tabakforsch. 12:251–258, 1984.
127. Sandler, D. P., R. B. Everson, and A. J. Wilcox. Passive smoking in adulthood and cancer risk. Am. J. Epidemiol. 121:37–48, 1985.
128. Sandler, D. P., R. B. Everson, A. J. Wilcox, and J. P. Browder. Cancer risk in adulthood from early life exposure to parents' smoking. Am. J. Public Health 75:487–492, 1985.
129. Sandler, D. P., A. J. Wilcox, and R. B. Everson. Cumulative effects of lifetime passive smoking on cancer risk. Lancet 1:312–315, 1985. (preliminary communication)

130. Schaefer, H. J. Comparative Evaluation of Exposure Distributions for Air Travelers and Radiation Workers. NAMRL-1159. Pensacola, Fla.: Naval Aerospace Medical Research Laboratory, 1972. (available from NTIS as AD-743482)
131. Schaefer, H. J. Radiation exposure on high-altitude passenger flights. Adv. Biol. Med. Phys. 13:153–206, 1970.
132. Schechter, M. S., and W. N. Sullivan. Gas-propelled aerosols and micronized dusts for control of insects in aircraft. II. Pesticide formulations. J. Econ. Entomol. 65:1444–1447, 1972.
133. Schievelbein, H., and F. Richter. The influence of passive smoking on the cardiovascular system. Prev. Med. 13:626–644, 1984.
134. Schmeltz, I., A. dePaolis, and D. Hoffmann. Chemical studies in tobacco smoke. 31. Phytosterols in tobacco: Quantitative analysis and fate in tobacco combustion. Beitr. Tabakforsch. 8:211–218, 1975.
135. Schoettlin, C. E., and E. Landau. Air pollution and asthmatic attacks in the Los Angeles area. Public Health Rep. 76:545–548, 1961.
136. Shabad, L. M., and G. A. Smirnov. Aircraft engines as a source of carcinogenic pollution of the environment [benzo(a)pyrene studies]. Atmos. Environ. 6:153–164, 1972.
137. Shephard, R. J., R. Collins, and F. Silverman. "Passive" exposure of asthmatic subjects to cigarette smoke. Environ. Res. 20:392–402, 1979.
138. Sheppard, D., W. L. Eschenbacher, H. A. Boushey, and R. A. Bethel. Magnitude of the interaction between the bronchomotor effects of sulfur dioxide and those of dry (cold) air. Am. Rev. Respir. Dis. 130:52–55, 1984.

139. Smeaton, M. J., P. J. Slater, and P. Robson. Malaria from a "commuter" mosquito . Lancet 1:845–846, 1984. (letter)
140. Speer, F. Tobacco and the nonsmoker: A study of subjective symptoms. Arch. Environ. Health 16:443–446, 1968.
141. Spendlove, J. C., and K. F. Fannin. Source, significance, and control of indoor microbial aerosols: Human health aspects. Public Health Rep. 98:229–244, 1983.
142. Spengler, J. D., D. W. Dockery, W. A. Turner, J. M. Wolfson, and B. G. Ferris, Jr. Long-term measurements of respirable sulfates and particles inside and outside homes. Atmos. Environ. 15:23–30, 1981.
143. Stahl, W. H., Ed. Compilation of Odor and Taste Threshold Values Data. DS48. Philadelphia, Pa.: American Society for Testing and Materials, 1973.
144. Stewart, A., and G. W. Kneale. Age-distribution of cancers caused by obstetric x-rays and their relevance to cancer latent periods. Lancet 2:4–8, 1970.
145. Stewart, A., and G. W. Kneale. Changes in the cancer risk associated with obstetric radiography. Lancet 1:104–107, 1968.
146. Stewart, A., J. Webb, and D. Hewitt. A survey of childhood malignancies. Br. Med. J. 1:1495–1508, 1958.
147. Sullivan, W. N., R. Pal, J. W. Wright, J. C. Azurin, R. Okamoto, J. U. McGuire, and R. M. Waters. Worldwide studies on aircraft disinsection at "blocks away." Bull. WHO 46:485–491, 1972.
148. Thayer, W. W. Tobacco smoke dilution recommendations for comfortable ventilation. ASHRAE Trans. 88(2):291–306, 1982.

149. Toxic and hazardous substances. Code of Federal Regulations, Title 29, Pt. 1910.1000. Washington, D.C.: U.S. Government Printing Office, 1985.
150. Trichopoulos, D., A. Kalandidi, L. Sparros, and B. MacMahon. Lung cancer and passive smoking. Int. J. Cancer 27:1–4, 1981.
151. U.S. Bureau of the Census. Statistical Abstract of the United States, 1985, 105th ed. Washington, D.C.: U.S. Government Printing Office, 1984.
152. U.S. Centers for Disease Control. Disinsection of aircraft. Federal Register 44(12 Oct.):58912, 1979.
153. U.S. Civil Aeronautics Board. Part 252-Smoking Aboard Aircraft: Final Rule. Amendment 1 to Part 252. Docket 29044. Washington, D.C.: U.S. Civil Aeronautics Board, 1983.
154. U.S. Environmental Protection Agency. Air Quality Criteria for Particulate Matter and Sulfur Oxides. EPA-600/8-82/029a. 3 vol. Washington, D.C.: U.S. Environmental Protection Agency, 1982.
155. U.S. Environmental Protection Agency. Revised Evaluation of Health Effects Associated with Carbon Monoxide Exposure: An Addendum to the 1979 EPA Air Quality Criteria Document for Carbon Monoxide. EPA-600/8-83/033F. Washington, D.C.: U.S. Environmental Protection Agency, 1984.
156. U.S. Environmental Protection Agency, Environmental Criteria and Assessment Office. Air Quality Criteria for Ozone and Other Photochemical Oxidants: External Review Draft No. 2. Vol. V. EPA-600/8-84-020B. Research Triangle Park, N.C.: U.S. Environmental Protection Agency, Environmental Criteria and Assessment Office, 1985.
157. U.S. Federal Aviation Administration. Aircraft cabin ozone contamination: Advance notice of proposed rulemaking. Federal Register 42(6 Oct.):54427–54428, 1977.

158. U.S. Federal Aviation Administration. FAA Aviation Forecasts, Fiscal Years 1985–1996. FAA-APO-85-2. Washington, D.C.: U.S. Federal Aviation Administration, 1985.
159. U.S. Federal Aviation Administration. Ozone Irritation During High Altitude Flight. Advisory Circular 00–52. Washington, D.C.: U.S. Federal Aviation Administration, 1977.
160. U.S. Federal Aviation Administration and U.S. National Institute for Occupational Safety and Health. Health Aspects of Smoking in Transport Aircraft. Cincinnati, Ohio: National Institute for Occupational Safety and Health, 1971. (available from NTIS as AD736097)
161. U.S. Surgeon General. Chronic Obstructive Lung Disease. DHHS (PHS) 84-50205. Washington, D.C.: U.S. Government Printing Office, 1984.
162. U.S. Surgeon General. The Health Consequences of Smoking: Cardiovascular Disease. Washington, D.C.: U. S. Government Printing Office, 1983.
163. Ventilation. Code of Federal Regulations, Title 14, Pt. 25.831. Washington, D.C.: U.S. Government Printing Office, 1985.
164. Vieillefond, H., P. Fourn, and R. Auffret. Characteristics in the atmosphere of long-range transport aircraft cabins. Aviat. Space Environ. Med. 48:503–507, 1977.
165. Wallace, R. W., and C. A. Sondhaus. Cosmic radiation exposure in subsonic air transport. Aviat. Space Environ. Med. 49:610–623, 1978.
166. Warhurst, D. C., C. F. Curtis, and G. B. White. A commuter mosquito's second bite? Lancet 1:1303, 1984. (letter)
167. Watkins, H. M. S. Epidemiologic investigations in Polaris submarines, pp. 9–53. In I. H. Silver, Ed. Aerobiology: Proceedings of the Third International Symposium on Aerobiology, University of Sussex, England, Sept. 1969. New York, N.Y.: Academic Press, 1970.

168. Weber, R. Travel involving long-distance flights: health risks. Med. Monatsschr. Pharm. 7 (5):147–151, 1984. (in German)
169. Weber, A., and T. Fischer. Passive smoking at work. Int. Arch. Occup. Environ. Health 47:209–221, 1980.
170. Weber, A., T. Fischer, and E. Grandjean. Passive smoking in experimental and field conditions. Environ. Res. 20:205–216, 1979.
171. Weber, A., T. Fischer, and E. Grandjean. Passive smoking: Irritating effects of the total smoke and the gas phase. Int. Arch. Occup. Environ. Health 43:183–193, 1979.

6

Health Effects Associated with Exposure to Airliner Cabin Air

The survey of airliner cabin contaminants in Chapter 5 suggests a diverse set of adverse health effects that could arise from exposure to the cabin environment—from acute effects, such as irritation, inflammation, and infection, to long-term effects, such as neoplasms, reproductive disorders, and decrement in pulmonary function. The following sections review the epidemiologic literature on adverse health effects that have stemmed from cabin air, as manifested in passengers and crews of commercial airliners. Where it is relevant, we also include studies on general aviation and military aircraft crews.

Our review of the literature overlaps with that of Kraus,[33] who reviewed epidemiologic studies of health effects in commercial pilots and flight attendants. Although his review focused on occupational effects unrelated to cabin air quality, he did present original data on occupational illness that are relevant to our study. He used 1979 California statistics on occupational illness and injury to compare the reported numbers of illnesses and injuries in flight attendants with the numbers expected, which were based on combined overall percentage distributions for all occupations. Flight attendants' reported occupational illness is generally much less than expected, except for infection, disease of the inner ear, respiratory disease, and aerotitis media; for these, ratios of observed to expected frequencies range from 9.8:1 (infections) to 209.1:1 (aerotitis media). It seems that the latter conditions are occupationally induced, but we know of no further relevant analyses of occupational-illness statistics.

We found almost no studies of health effects in airliner passengers, other than a few isolated case reports of miscellaneous diseases. The one exception is a Danish retrospective study[47] of 773 airline passengers admitted to a Danish emergency ward from Copenhagen Airport in Kastrup in 1975–1976. The estimated annual turnover at that airport during the period was 8–9 million passengers. The most common illnesses were injury and poisoning (219 passengers), "symptoms, signs, and ill-defined conditions" (120), diseases of the circulatory system (67), diseases of the respiratory system (62), and infections and parasitic diseases (57). Many illnesses classed as "symptoms, signs, and ill-defined conditions" might have been related to the cabin environment, including hyperventilation, syncope and collapse, and ear problems; but it is not possible to attribute any of these illnesses or injuries directly to cabin air quality.

HEALTH EFFECTS OF CONCERN

Irritation and Inflammation

Passengers and cabin crew in an airliner can be exposed to a number of substances that can cause eye, nose, and respiratory irritation, which also appear to be commonly reported by passengers and crew when complaining about the air quality in airplanes. There has been relatively little evaluation of these symptoms among aircraft occupants.

A questionnaire survey of flight attendants in 1978 found a high prevalence of reported eye discomfort.[21] [22] The survey form was distributed through the monthly magazine of the Association of Flight Attendants, which at that time represented flight attendants on 18 major airlines. Of the 774 who responded, 95% reported some eye discomfort while on an aircraft. Dry eyes and redness were reported by approximately 90% of the respondents; fewer reported other eye symptoms, such as blinking, blurred vision, and tearing. Over 90% reported smoking as a cause of their discomfort. Air-conditioning, cabin lights, wing reflection, and napping were also reported as contributing to eye problems. In general, the complaints were not correlated with the use of contact lenses, but attendants wearing soft contact lenses did report more problems with blurred vision and tearing. Although a high prevalence of eye discomfort was reported, the value of the study's conclusions is limited by the possibility of selection bias in the respondents to the survey and by the lack of a comparison group.

Three studies have attempted to evaluate symptoms due to ozone exposure on aircraft. In 1978, Reed et al., at the California State Department of Health, conducted a questionnaire study of flight attendants in three airlines: Pan American World Airways, which usually flew long distances at high altitudes; Pacific Southwest Airlines, which flew only short distances at lower altitudes; and Trans World Airlines, which flew both types of routes.[50]

Of questionnaires mailed to 3,280 flight attendants, 1,330 completed questionnaires were received. The authors estimated that 61% of flight attendants on active status returned questionnaires, which included questions on symptoms related to ozone exposure, on other risk factors, on characteristics of the flights, and on the time course of symptoms. Ozone exposures were believed to be much higher in the high-altitude long-distance flights. The prevalence of some symptoms possibly related to ozone exposure (chest pain, difficulty in breathing, and persistent cough) was significantly higher among the attendants on the high-altitude flights than among those in the other two airlines. No significant differences were found in the prevalence of other symptoms, such as extreme fatigue and back pain, which would not be expected to be caused by exposure to ozone. Flying on particular high-altitude aircraft (e.g., B-747 and B-747-SP, not equipped with a catalytic unit to abate ozone in the cabin air) was associated with symptoms of ozone toxicity. In general, this study found a higher prevalence of symptoms of ozone toxicity among flight attendants with higher exposures to ozone. Although limited by the response rate and the lack of direct ozone measurements, this study did indicate possible problems due to exposures during high-altitude long-distance flights.

Another questionnaire study of symptoms due to ozone exposure among flight attendants was conducted in 1977 by Tashkin et al., at the University of California, Los Angeles.[52] A questionnaire directed at flights on the B-747-SP was sent to 450 flight attendants in the Los

Angeles area, of whom 248 responded; a questionnaire directed at flights on the B-747 was sent to the same 248 attendants and produced only 38 responses; and a similar questionnaire directed at both aircraft was sent to 850 attendants in the New York area, of whom only 65 responded. The questionnaire results were evaluated by three independent observers, who knew which aircraft the respondents had worked on and who graded symptoms on the basis of their possible relationship to ozone exposure. In addition, 21 flight attendants who had experienced severe respiratory symptoms while on B-747-SP aircraft received a more detailed medical evaluation, including pulmonary function testing, about 2 wk after the problem flight. The attendants who flew on the B-747-SP aircraft reported a higher prevalence of ozone-related symptoms (throat irritation, cough, difficulty in breathing, etc.) while on the aircraft than afterward and a higher prevalence than the attendants who flew on the standard B-747 in both the New York and Los Angeles portions of the study. The results of all the pulmonary function testing on the 21 selected participants were normal, as expected on the basis of time since exposure. Although it suggested that symptoms were due to ozone exposure, this study was limited by the lack of direct measurements of ozone exposure and by the very poor rate of response to the questionnaires, particularly in the comparison group.

A third study was performed by the Occupational Health Clinic of San Francisco General Hospital in 1984 at the request of the Independent Union of Flight Attendants.[30] The study was designed to see whether the prevalence of ozone-related symptoms reported by Reed et al.[50] persisted on long-haul, high-altitude flights and whether respiratory symptoms on these flights were associated with objective decreases in pulmonary function. The study consisted of two phases. In Phase I, all Pan American World Airways flight attendants based in San Francisco, London, or California were mailed a self-administered questionnaire concerning symptoms, medical diagnoses, smoking, and occupational history. Results of the questionnaire were compared with the data of Reed et al. A small selected group of flight attendants who noted ozone-related symptoms on the questionnaire were asked to participate in Phase II. For Phase II, each participant was instructed in the use of a Mini-Wright peak flow meter to measure peak expiratory flow rate (PEFR) and measured PEFR every 2 h while awake (total duration was not reported). Preflight (more than 12 h since last flight) and postflight (any time within 12 h of landing) PEFR measurements were then compared with unpaired t tests.

For Phase I, approximately 1,000 flight attendants were sent the survey; 280 returned completed questionnaires. A followup survey indicated that many nonrespondents failed to respond because their identity would not be protected. Demographically, the responding sample was similar to that of the Reed et al. sample, but older. Prevalence rates of chest pain or tightness (65%), shortness of breath (65%), and cough (57%) were similar to or slightly higher than those reported by Reed et al. Symptoms were more prevalent on B-747-SP flights and were more prevalent among those who had ever smoked than among nonsmokers.

Of the 20 flight attendants asked to participate in Phase II, only eight yielded analyzable data. Mean preflight PEFRs were always higher than postflight PEFRs, by 7-35 L/min (average, 21 L/min). The statistical analysis of the data is incorrect, so it is difficult to judge the statistical significance of these results. Two flight attendants had preflight-postflight differences in PEFR of over 20% in 24 h associated with long flights.

These results suggest that efforts to reduce onboard ozone concentrations have not had an effect on the prevalence of ozone symptoms and that flights might be accompanied by decreases in PEFR. Phase I had several limitations, including a lower response rate and an older population than the Reed et al. study, a self-administered questionnaire, and a lack of ozone measurements. Phase II was hampered by very small numbers, self-selection, use of flow meters with questionable accuracy, self-administered data collection, an ambiguous protocol for data collection (which allowed different persons to contribute different numbers of observations), a lack of ozone measurements in flight, and inappropriate data analysis. If these limitations are kept in mind, this study's conclusions can be regarded as only suggestive until confirmed by appropriately designed studies.

In each instance of potential irritation and inflammation, passengers and crews with pre-existing disease or disorders of the organs affected suffer increased effects. People with upper respiratory infections suffer more from pressure changes and possibly from low humidity. People wearing soft contact lenses have more eye symptoms that result from low humidity. Patients with chronic pulmonary disease might have more symptoms from inhaling ozone. The medical literature discusses these increased susceptibilities, but does not document them.

Asymptomatic sinus disease has been the subject of a number of studies. One study[23] of 211 Air Force pilots aged 25–35 showed radiographic evidence of maxillary sinus abnormality in 25%, but no control group was studied. A followup study compared these Air Force pilots with two groups of Air Force employees who had no flying experience. One comparison group consisted of 100 new airmen trainees who were below age 25; the second consisted of 100 men aged 25–35 who were patients in an Air Force hospital for diagnostic procedures not related to ear, nose, or throat symptoms and had no flying experience. The prevalence of maxillary sinus abnormality among the two control groups was 26% and 29%, respectively. Another study[27] compared 1,284 asymptomatic flyers with a control group of 200 nonflyers. The reported prevalence of abnormalities of the paranasal sinuses was 22% in the control group and 15.6% in the flyers. Selection of the controls and comparability of the two groups were not reported.

Other conditions associated with mucous membrane inflammation have been found in airliner cabin occupants. Aerotitis media and other middle ear conditions have been reported as significant health problems for flight attendants.[33] [55] These conditions might be due to cabin pressure changes, but mucous membrane inflammation could contribute to them.

Infection

Only one study has clearly documented the occurrence of an outbreak of infectious disease related to airplane use.[43] An outbreak of influenza occurred in 1978 in Alaska. Because of an engine malfunction, an airliner with 54 persons aboard was delayed on the ground for 3 h, during which the aircraft ventilation system was reportedly turned off. Within 3 d of the incident, 72% of the passengers became ill with influenza. One passenger (the index case) was ill while the aircraft was delayed. Serologic evidence of influenza infection was found in 20 of 22 passengers tested, and the virus was isolated from eight of 31 passengers whose serum was cultured. Documentation of this outbreak was assisted by the circumstance that all the passengers traveled to one small town and by the alertness of the local physician. Similar outbreaks could result from crowded flights with an infectious person and not be documented or noticed, because passengers would disperse after landing.

Persons with coincidental acute and chronic infections suffer more from superimposed infections acquired on the aircraft. In addition, increasing numbers of people with diminished resistance to infection might be traveling as passengers—specifically, patients undergoing chemotherapy or x-ray therapy for malignancies and those infected with the HTLV-III virus (acquired immune deficiency syndrome). There is no evidence that that virus can be transmitted through the air.

Respiratory Impairment

Various constituents of the aircraft environment could lead to respiratory impairment in passengers or crew. The manifestations of respiratory impairment are diverse and include pulmonary diseases, acute respiratory illness, sinus disease, sarcoidosis, and spontaneous pneumothorax.

Several studies have investigated pulmonary function in flight attendants or pilots. One report found that higher percentages of members of self-selected groups of Miami-and New York-based Pan American World Airways flight attendants, but not San Francisco-based flight attendants, had spirometric abnormalities than of an age-and sex-matched Michigan group.[44] The finding is difficult to interpret, because of the self-selection process, questions of comparability of measurements in the flight attendants and the Michigan group, and failure to take smoking history into account. Another study reported, as expected, an absence of pulmonary function abnormalities in a select group of 21 flight attendants who were tested 2 wk after experiencing respiratory symptoms during B-747-SP flights.[52]

A study of 257 active United Airlines pilots revealed that 12% had evidence of minimal to moderate ventilatory impairment.[15] Disease prevalence increased with age and smoking history, but no comparisons were made with a nonpilot population, so it is difficult to assess the importance of the finding. Similar findings have been reported for general aviation airmen.[37]

Dille[20] compared the prevalence of asthma, emphysema, bronchiectasis, bronchitis, and other unclassified pulmonary diseases in a population of 288,000 active civil airmen with the prevalence of these diseases reported in the U.S. National Health Survey and found a much higher prevalence in the general population. That was expected, because of the self-selection of active airmen. The long-term followup of the U.S. Navy's "1,000-aviator cohort" revealed that decrements in pulmonary function were associated with cigarette-smoking, coronary arterial disease, and weight gain.[38] No correlation was reported between a career in military aviation and the development of pulmonary disease; but a career in military aviation was a dichotomous variable, which was coded (present) if a person had 15 yr or more of flying history and not coded (absent) otherwise, and is at best a weak measure of exposure.

Several incidental reports have noted spontaneous pneumothorax in pilots, but presented no comparisons with nonpilot populations, so it is impossible to judge whether the risk is increased by a flying career or by onboard environmental conditions.[19 24 25]

One British investigation[11] has studied the prevalence of pulmonary lesions resembling sarcoid granulomata in 2,000 autopsy reports after 700 aviation accidents. Military crews had a higher rate than civil airmen, who had a higher rate than passengers or glider pilots. Review of the incidence of clinical sarcoidosis in the Royal Air Force in 1962–1977 showed a much greater overall incidence than the 3 per 100,000 in the general

U.K. population; the aircrew incidence averaged 14.4 cases per 100,000, and the ground crew, 10.8 cases per 100,000. Because of inconsistencies in the autopsy reports and clinical incidence rates and the lack of corroborating evidence, no conclusions were drawn by the author. This pathologic but often asymptomatic lesion needs to be searched for in other well-controlled studies.

Jasinski[31] showed acute respiratory illness to be a common problem in flight attendants, but it cannot be determined from the report whether the incidence was higher than that found in other populations. An Italian study[48] of morbidity in flying personnel appeared to suggest higher rates of acute respiratory illness than in nonflying airline employees, but the details of the study were not reported. The work by Kraus[33] cited earlier suggested higher rates of respiratory disease in flight attendants.

Isolated autopsy findings of hypoxia, intoxication, hyperventilation, and carbon monoxide intoxication in military pilots have been reported.[26 36 49] The relevance of these reports to the commercial airliner cabin environment is uncertain. One report[42] showed that contamination of the ventilation system (in military aircraft) with lubricating oil could lead to intoxication.

As noted earlier, patients with underlying pulmonary disease are more susceptible to changes in cabin air that affect pulmonary function. Thus, any increase in the partial pressure of carbon dioxide (pCO_2) in the air will adversely affect patients with chronic obstructive pulmonary disease (COPD) who are already functioning with an increased blood pCO_2 and increased alveolar pCO_2. A further increase might make it even more difficult to maintain a normal blood pH. Similarly, the decrease in oxygen partial pressure (pO_2) that occurs at 8,000 ft is safe for normal people, but possibly hazardous for patients with COPD. One study of patients with COPD who were placed at reduced pO_2 as they would be at 8,000 ft showed that patients without overt pulmonary failure can expect no trouble and that those with symptomatic COPD can be tested in advance by their physicians to determine whether they will need supplementary oxygen if they must fly.[51]

Cardiovascular Effects

The effects of cabin air quality on cardiovascular function in normal persons and patients with underlying disease are of interest. There is no evidence of any effects in people with normal hearts and blood vessels, other than occasional anecdotes of venous thrombosis and pulmonary embolism, which are much more likely to be associated with inactivity than with air quality. A high percentage of adults have some underlying coronary arterial disease, which theoretically could be made worse by the products of cigarette-smoking.[3 7–9] Although angina pectoris might result from myocardial ischemia, there is no evidence that myocardial infarction would be caused by inadequacies in cabin air quality.

The many papers on coronary arterial disease and resulting sudden death of pilots are not reviewed here, because they are concerned with screening and related health examinations, rather than with possible deleterious effects of cabin air.

Some persons with symptomatic cardiovascular disease are under medical care, so decisions about the possibly increased hazards of reduced pO_2 and exposure to cigarette smoke, carbon monoxide, and ozone could be made by their physicians.

Neoplasms

Several constituents of cabin air might increase the risk of neoplasia, including passive smoking and exposure to radiation. However, published reports contain little documentation of cancer incidence in flying personnel.

Kraus[33] reviewed Milham's study[41] of occupational mortality in Washington State, which gave proportional mortality ratios for many occupational categories, including pilots, navigators, and flight attendants. Statistically significant increases were seen in rectal cancer in pilots and navigators, but significant reductions in lung cancer. These observations have not been confirmed in other data bases.

In 1981, the Centers for Disease Control carried out a health hazard evaluation for the Independent Union of

Flight Attendants (IUFA).[45] IUFA mailed a questionnaire to approximately 6,000 of its members. Responses were received from 9%; the reason for the low response rate is that a response was requested only if the member had cancer. Crude incidence and prevalence rates were compared with statistics from the Birmingham Regional and Connecticut Tumor Registries. Only skin cancer showed an excess risk among flight attendants: 3–10 times the expected rates. The possible environmental causes relevant to skin cancer are exposures to sunlight, ionizing radiation, arsenicals, and hydrocarbons. Although the results were suggestive, the study had clear limitations, including the possible failure of those who might have had cancer to respond, unconfirmed self-reported diagnoses, and lack of a control group.

Three case reports of nasopharyngeal cancer in bush pilots suggested that the cancers could be related to pressure changes,[5 6 54] but presented no substantiating evidence.

Reproductive Disorders

A few studies on reproductive disorders or pregnancy outcome in flight attendants have been reported. Menstrual disorders are thought to be related primarily to stress and interruption of circadian rhythm, but there is some speculation that they can be attributed to solar radiation,[16] which might also predispose to unfavorable pregnancy outcome.

Cameron[14] presented data on menstrual function in 98 Swiss flight attendants. Long-term followup data on reproductive outcome were available on only 50 women. There was a general suggestion that no increase in menstrual disorders was associated with flying, but that the miscarriage rate among married ex-hostesses was high. Iglesias et al.[29] reported the results of interviews of 200 flight attendants who sought medical assistance for various clinical problems; 39% reported unfavorable changes in menstrual cycles 6–24 mo after beginning aeronautical service. Both these studies had problems of recall and self-reporting, lack of controls, self-selection, and small numbers of participants (particularly the long-term followup in the Cameron study), so no reliable conclusions can be drawn.

A Czechoslovakian study[28] found a significantly higher percentage of pathologic pregnancies and deliveries in flight attendants after 2–7 yr of flying than in the general population. High rates of spontaneous abortion and premature delivery were also reported. Details of this study were not available to the Committee, but it does not appear to have used appropriate controls.

One study[34] compared pregnancy outcome in U.S. Air Force women with age-and time-matched civilian patients. Although the Air Force women had generally higher rates of perinatal death, low-birthweight babies, small-for-gestational-age babies, and prematurity, the differences were not statistically significant; there were several significant differences in risk factors, including nulliparity, race, and marital status. In addition, the duties of the Air Force women in this study varied, so they were not good surrogates for flight attendants.

Miscellaneous

Mendez Martin[40] surveyed various Spanish studies that showed that urinary calculosis was a common disease of flight personnel, possibly attributable in part to their low-humidity environment. English-language reports on this problem are sparse.[32]

Summary

The available information on the health of airliner crews and passengers stems largely from ad hoc epidemiologic studies or case reports of specific health outcomes, although occupational-health statistics have been used in at least one study. The conclusions that can be drawn from the available data are limited to a great extent by the self-selection of the subjects of studies, the lack of comparison groups, and a lack of exposure information. The major findings must be reviewed with this caveat in mind.

The one study that used occupational-health statistics[33] found that flight attendants had higher rates of respiratory disease, aerotitis media, infections, and diseases of the inner ear than other

California workers. Although these findings are important, the Committee feels that they should be verified by using additional occupational-health statistics from different sources and periods. The lack of specific exposure information makes it difficult to attribute the high rates to cabin air quality. However, increased rates of aerotitis media in flight personnel have been documented in other studies.[55]

A higher prevalence of ozone-related (self-reported) symptoms was found in flight attendants on long, high-altitude flights than on short, low-altitude flights.[50] Despite some limitations in the study, it offered some evidence that ozone-related health problems exist among flight attendants. Results of another study[30] suggested that the prevalence of ozone-related symptoms continues to be high. Neither study correlated reported symptoms with direct onboard ozone measurements. To our knowledge, there have been no similar studies on passengers.

One epidemiologic study[43] documented that outbreaks of influenza can be associated with unusual operating conditions, but the incidence of such outbreaks is unknown, as is their dependence on operating conditions.

The literature on respiratory disease is sparse and fragmented and is of no value in assessing health effects associated with cabin air quality. Except for miscellaneous reports, there is no solid information on an association of neoplasia with cabin air quality.

The English-language reports on pregnancy outcome in flight attendants are flawed, and the Committee has not fully evaluated foreign-language reports that purport to show increased rates of unfavorable pregnancy outcome. The effect of cabin air quality in inducing unfavorable pregnancy outcome is also unknown.

Mendez Martin[40] has reported that urinary calculosis is common in flight personnel, but the Committee is aware of no corroboration of this finding.

MONITORING AND SURVEILLANCE OF CREW AND PASSENGER HEALTH

Data on the health of passengers and crew have three potential sources: airlines, flight attendant unions, and FAA. The populations monitored can be conveniently divided into pilots, flight attendants, and passengers.

Pilots

FAA requires medical certification of pilots every 6 mo and thus has considerable information on the health status of active civil airmen,[12] as well as statistics on medical disqualifications.[4] [18] Because of this requirement and the expense of training pilots, many airlines routinely monitor the health status of their pilots.[46]

Several studies have looked at the health of pilots. Buley[13] and Kulak et al.[35] Investigated cases of in-flight airline incapacitation, primarily with an eye to correlating such incidents with accidents and to determining whether stricter medical certification could reduce in-flight incapacitation. No attempt was made to relate incapacitation to specific occupational hazards or to contrast incidence rates with those in a comparison group. There has been one long-term followup study of mortality and morbidity in military pilots.[38] As one might expect, their health is better than that of the general population or of age-matched Framingham men, but there was no comparison with a suitably selected control group.

The extensive data available on the health of pilots are of little use in studying the health effects of cabin air. A primary limitation is that the special cockpit environment is not indicative of the general cabin environment. In addition, the orientation of health monitoring is to ensure that certified pilots are free of health problems that might jeopardize their ability to operate a plane safely; thus, its purpose is not to detect potential health effects of the working environment. For the system to meet the latter purpose, several important and fundamental changes would need to be made, including the addition of followup of retired airmen, elimination of self-selection problems (airmen who, for health reasons, elect not to renew their licenses do not appear in the current records), collection of additional data (on both health and exposure) pertinent to occupational hazards, and the implementation of a sophisticated statistical analysis and reporting system.

Flight Attendants and Passengers

No monitoring or surveillance activities appear to be directed solely at the health of flight attendants. A few airlines indicated that some pre-employment health data were available to them and that some additional medical records on selected flight attendants were kept. However, these records are considered proprietary and were not available to the Committee. More important, it appears that no airline monitors the health of all its flight attendants routinely. Airlines do maintain records of workers' compensation and disability claims, but only a portion of these data can be released. A few airlines appear to keep records of employees' service histories (flight times, routes, and types of service).

A few airlines indicated that they maintain records on incidents of passenger illness (some limited only to oxygen use and passenger complaints about air quality), but the adequacy of these records for monitoring purposes is unknown.

Other than accident and incident data (see the following section), FAA collects no data on the health of flight attendants or passengers.

The flight attendant unions have periodically sponsored mail-questionnaire surveys on health-related issues, but do not sponsor routine data collection directed at monitoring the health of flight attendants.

The Association of Flight Attendants receives reports submitted by flight attendants concerning poor cabin air quality. From January 1977 to April 1982, 297 reports were received, and descriptive statistics were tabulated for presentation to the Subcommittee on Aviation of the U.S. Senate Committee on Commerce, Science, and Transportation. The value of these reports in assessing health risks is questionable, in that they appear to be voluntary and therefore self-selected. In addition, no standard protocol for reporting is used, so the information gathered is fragmentary and selective. The number of incidents reported per year from 1977 through 1982 varied erratically (21, 70, 6, 46, 135, and 66). In view of the number of flights per year, the reported incident rates seem low, although there might be some underreporting; there is no basis on which to establish an expected rate for these reports.[10]

FAA Surveillance Activities

FAA has claimed regulatory jurisdiction over the cabin as a workplace. FAA asserts that its responsibility toward passengers is related to their safety and claims not to have regulatory authority over health. No federal agency monitors the health of flight attendants.

Other than the medical data collected for pilot certification (discussed above), the health data on passengers or flight crews that are systematically collected by FAA are very limited. They are reported in the Accident/Incident Data System (AIDS), described as follows in the AIDS user's guide:[53]

> The Accident/Incident Data System (AIDS) contains data records for general aviation accidents/incidents, air carrier incidents, and, beginning with 1982, air carrier accidents. The system consists of various data bases, computer hardware, computer programs and manual procedures which in combination produce a functional capability for the user. The system gives additional data elements, provides for English-like retrievals and reports, puts emphasis on ad hoc retrievals, provides easily utilized standard reports and provides user access through data terminals.
>
> The basic design of the AIDS system is to provide the user with current and accurate information about general aviation accidents and incidents coupled with the facility to produce "standard" reports and "ad hoc" reports based on specific requirements. Several standard report

> formats can be requested by specifying: time-period of interest, national/regional criteria, and event selection criteria (type of accident, etc.). Specialized queries can be prepared and input by trained users. Additionally, the tools exist for conducting statistical analyses of the data contained in the data base.

The objectives of AIDS are laudable, and the Committee is optimistic that the system will prove to be a valuable research aid for aviation safety. We were impressed by the documentation of the computer system designed to gain access to the data base and by the data coding system.

However, AIDS is relatively new and has yet to realize its full potential. The Committee experienced two difficulties in using the system. First, we were unable to find an accessible, concise, and thorough description of the collection system and its data base contents. Without good information on the data collection process, the Committee found it difficult to judge the quality of the data (for health monitoring purposes) and the desirability of using them for health monitoring. For example, the description of the criteria used for defining an accident or an incident in the FARs[39] is insufficient to enable one to be certain of the quality of health data that enter AIDS. In addition, the Committee found that, although access to the data base itself appears good, descriptive and summary statistics on such items as number of fires by cause and number of passenger deaths by cause were not readily accessible.

In summary, the Committee feels that AIDS has potential as a health monitoring or surveillance tool, but that considerably more effort by FAA will be required to make it effective. It is insufficient merely to collect data and provide access to them. It is important that at least basic statistical summaries of key information be produced routinely. The Committee also notes that the purpose of AIDS is to monitor accidents and incidents; therefore, in its current form it has no value for monitoring chronic health effects of air travel in passengers or crew.

GROUPS AT INCREASED RISK

Aircraft at cruising altitudes maintain artificial cabin altitudes of 5,000–8,000 ft. Because of the associated decrease in pO2 compared with that at sea level, passengers with specific health problems might be at increased risk while flying. A number of committees in special and general medical associations publish guidelines for physicians to use in advising patients about air travel.[1 17] Of most general coverage is a list, prepared by the American Medical Association's Commission on Emergency Medical Services, of conditions in which air travel is contraindicated.[2] The list is presented here for information, although the Committee found little material on these conditions in passengers traveling on aircraft.

- Cardiovascular—myocardial infarction within the preceding 4 wk, cerebrovascular accident within the preceding 2 wk, severe hypertension, decompensated cardiovascular disease, or any condition that restricts cardiac reserve. Patients with chronic cardiovascular problems, such as cyanotic congenital heart disease or coronary insufficiency, should have supplemental oxygen whenever flight altitude is greater than 22,500 ft.
- Bronchopulmonary—pneumothorax, congenital pulmonary anomaly, or vital capacity less than 50%. Patients with chronic pulmonary problems —such as cystic fibrosis, emphysema, chronic asthma, or fibrotic pulmonary conditions—should have supplemental oxygen whenever flight altitude is greater than 22,500 ft.
- Eye, ear, nose, and throat—recent eye surgery, acute sinusitis, or acute otitis media. Patients who must fly during the congestive stage of upper respiratory infection should use local shrinking agents or oral decongestants.
- Gastrointestinal—abdominal surgery within the preceding 2 wk, acute diverticulitis or ulcerative colitis, acute esophageal varices, or acute gastroenteritis.

- Neuropsychiatric—epilepsy (unless it is well controlled medically and simulated cabin altitude is never greater than 8,000 ft), recent skull fracture, brain tumor, or history of violent or unpredictable behavior.
- Hematologic—anemia (hemoglobin concentration of less than 8.5 g/dL or red-cell count of less than 3 million/mm3 in adults), sickle-cell disease (unless cruising altitude is never greater than 22,500 ft), or hemophilia.
- Pregnancy—beyond 240 d or if miscarriage is threatened.
- Miscellaneous—Scuba divers should not fly for at least 12 h after diving—24 h after repeated deep diving—before flying. The flight surgeon should be consulted if a patient requires intravenous fluids or special medical apparatus.

REFERENCES

1. Aerospace Medical Association, Committee on Medical Criteria. Medical criteria for passenger flying. Arch. Environ. Health 2:124–138, 1961.
2. American Medical Association, Commission on Emergency Medical Services. Medical aspects of transportation aboard commercial aircraft. JAMA 247:1007–1011, 1982.
3. Anderson, E. W., R. J. Andelman, J. M. Strauch, N. J. Fortuin, and J. H. Knelson. Effect of low-level carbon monoxide exposure on onset and duration of angina pectoris: A study in ten patients with ischemic heart disease. Ann. Intern. Med. 79:46–50, 1973.
4. Anderson, R., and C. C. Gullett. Airline pilot disability: Economic impact of an airline preventive medicine program. Aviat. Space Environ. Med. 53:398–402, 1982.
5. Andrews, P. A., and L. Michaels. Aviator's cancer. Lancet 2:640, 1968. (letter)

6. Andrews, P. A. J., and L. Michaels. Nasopharyngeal carcinoma in Canadian bush pilots. Lancet 2:85–87, 1968.
7. Aronow, W. S. Aggravation of angina pectoris by two percent carboxyhemoglobin. Am. Heart J. 101:154–157, 1981.
8. Aronow, W. S. Effect of carbon monoxide on cardiovascular disease. Prev. Med. 8:271–278, 1979.
9. Aronow, W. S. Effect of passive smoking on angina pectoris. N. Engl. J. Med. 299:21–24, 1978.
10. Association of Flight Attendants. Exhibit A (passenger comments and flight attendant symptoms, 01/77–01/82). In U.S. Senate, Committee on Commerce, Science, and Transportation, Subcommittee on Aviation (97th Congress, 2nd Session). Airliner Cabin Safety and Health Standards: Hearing on S. 1770, May 20, 1982. Washington, D.C.: U.S. Government Printing Office, 1982.
11. Balfour, A. J. Sarcoidosis in aircrew. Aviat. Space Environ. Med. 53:269–272, 1982.
12. Booze, C. F., Jr. Health Examination Findings Among Active Civil Airmen. FAA-AM-84-8. Oklahoma City, Okla.: U.S. Federal Aviation Administration, Civil Aeromedical Institute, 1984.
13. Buley, L. E. Incidence, causes and results of airline pilot incapacitation while on duty. Aerosp. Med. 40:64–70, 1969.
14. Cameron, R. G. Effect of flying on the menstrual function of air hostesses . Aerosp. Med. 40:1020–1023, 1969.
15. Catlett, G. F., and G. J. Kidera. Detection of respiratory impairment in pilots. Aerosp. Med. 40:1251–1252, 1969.

16. Cope, F. W. Idiopathic menstrual disorders in airline stewardesses: A possible origin from solar radiation of heavy magnetic particles. Int. J. Biometeorol. 25:219–221, 1981.
17. Cystic Fibrosis Foundation, Cardiovascular Committee. Airline travel for children with chronic pulmonary disease. Pediatrics 47:408–410, 1976.
18. Dark, S. J. Medically Disqualified Airline Pilots. FAA-AM-84-9. Oklahoma City, Okla.: U.S. Federal Aviation Administration, Civil Aeromedical Institute, 1984.
19. Dermksian, G., and L. E. Lamb. Spontaneous pneumothorax in apparently healthy flying personnel. Ann. Intern. Med. 51:39–51, 1959.
20. Dille, J. R. Pulmonary disease in general aviation. Aerosp. Med. 37:732–735, 1966.
21. Eng, W. G. Survey on eye comfort in aircraft. I. Flight attendants. Aviat. Space Environ. Med. 50:401–404, 1979.
22. Eng, W. G. Survey on eye comfort in aircraft. II. Use of ophthalmic solutions. Aviat. Space Environ. Med. 50:1166–1169, 1979.
23. Fascenelli, F. W. Maxillary sinus abnormalities. Arch. Otolaryngol. 90:190–193, 1969.
24. Flux, M., and J. R. Dille. Inflight spontaneous pneumothorax: A case report. Aerosp. Med. 40:660–662, 1969.
25. Fuchs, H. S. Incidence of spontaneous pneumothorax in apparently health aircrews. Aerosp. Med. 38:1286–1288, 1967.
26. Gibson, T. M. Hyperventilation in aircrew: A review. Aviat. Space Environ. Med. 50:725–733, 1979.
27. Hanna, H. H. Asymptomatic sinus disease in aircrew members. Aerosp. Med. 45:77–81, 1974.

28. Hinst, J., and D. Bruchac. The influence of civil aviation upon the reproductive system of the woman. Cesk. Gynekol. 46:456–459, 1981. (in Czech)
29. Iglesias, R., A. Terres, and A. Chavarria. Disorders of the menstrual cycle in airline stewardesses. Aviat. Space Environ. Med. 51:518–520, 1980.
30. Independent Union of Flight Attendants. Flight Attendants Health Survey: Interim Report #1. San Francisco, Cal.: San Francisco General Hospital, Northern California Occupational Health Center, 1985.
31. Jasinski, D. R. Some health problems of women flight attendants in the Pacific. Hawaii Med. J. 39:39–42, 1980.
32. Kohler, F. P. Asymptomatic upper urinary tract lithiasis in flying personnel. J. Urol. 86:370–372, 1961.
33. Kraus, J. F. Epidemiological studies of health effects in commercial pilots and flight attendants: A review. Sangyo Ika Daigaku Zasshi 7(Suppl.):32–44, 1985.
34. Kruger, P. S. Risk factors and pregnancy outcome among Air Force women. Milit. Med. 144:788–791, 1979.
35. Kulak, L. L., R. L. Wick, Jr., and C. E. Billings. Epidemiological study of in-flight airline pilot incapacitation. Aerosp. Med. 42:670–672, 1971.
36. Lacefield, D. J., P. A. Roberts, and P. M. Grape. Carbon Monoxide In-Flight Incapacitation: An Occasional Toxic Problem in Aviation. FAA-AM-82-15. Oklahoma City, Okla.: U.S. Federal Aviation Administration, Civil Aeromedical Institute, 1982.

37. Lategola, M. T., M. Flux, and P. J. Lyne. Spirometric assessment of potential respiratory impairment in general aviation airmen. Aviat. Space Environ. Med. 48:508–511, 1977.
38. MacIntyre, N. R., R. E. Mitchell, A. Oberman, W. R. Harlan, and A. Graybiel. Long-term follow-up of lung volume measurements in initially healthy young aviators. Aviat. Space Environ. Med. 52:1–5, 1981.
39. Mechanical reliability reports. Code of Federal Regulations, Title 14, Pt. 121.703. Washington, D.C.: U.S. Government Printing Office, 1985.
40. Mendez Martin, A. Urolithiasis among flight personnel. Med. Clin. (Barcelona):72:65–70, 1979. (in Spanish)
41. Milham, S., Jr. Occupational Mortality in Washington State, 1950–1979. NIOSH 83-116. Cincinnati, Ohio: National Institute for Occupational Safety and Health, 1983. (available from NTIS as PB-84-199769)
42. Montgomery, M. R., G. T. Wier, F. J. Zieve, and M. W. Anders. Human intoxication following inhalation exposure to synthetic jet lubricating oil. Clin. Toxicol. 11:423–426, 1977.
43. Moser, M. R., T. R. Bender, H. S. Margolis, G. R. Noble, A. P. Kendal, and D. G. Ritter. An outbreak of influenza aboard a commercial airliner. Am. J. Epidemiol. 110:1–6, 1979.
44. National Institute for Occupational Safety and Health. Flight Attendants Health Survey: Interim Report #1. HETA 81-331. Cincinnati, Ohio: National Institute for Occupational Safety and Health, 1982.
45. National Institute for Occupational Safety and Health. Independent Union of Flight Attendants: Interim Report #1. HETA 82-083. Cincinnati, Ohio: National Institute for Occupational Safety and Health, 1982.

46. Orford, R. B. A critical review of the periodic examination of pilots at the Mayo Clinic 1939–1973 to determine the effectiveness of a comprehensive health maintenance program in an airline pilot population. (unpublished thesis, University of Minnesota, 1975)
47. Pilgaard, S., A. Koster, and D. Munck. Airline passengers admitted to an emergency ward: A 2-year follow-up study. Ugreskr. Laeger 141:499–502, 1979. (in Danish)
48. Polizzi di Sorrentino, A. Influence of flight and commercial piloting on morbidity of civil aviation personnel. Riv. Med. Aeronaut. Spaz. 32:42–62, 1969. (in Italian)
49. Rayman, R. B., and G. B. McNaughton. Hypoxia: USAF experience, 1970–1980. Aviat. Space Environ. Med. 54:357–359, 1983.
50. Reed, D., S. Glaser, and J. Kaldor. Ozone toxicity symptoms among flight attendants. Am. J. Ind. Med. 1:43–54, 1980.
51. Schwartz, J. S., H. Z. Bencowitz, and K. M. Moser. Air travel hypoxemia with chronic obstructive pulmonary disease. Ann. Intern. Med. 100:473–477, 1984.
52. Tashkin, D. P., A. H. Coulson, M. S. Simmons, and G. H. Spivey. Respiratory symptoms of flight attendants during high-altitude flight: Possible relation to cabin ozone exposure. Int. Arch. Occup. Environ. Health 52:117–137, 1983.
53. U.S. Federal Aviation Administration, Flight Standards National Field Office. Accident/Incident Data System—AIDS—User Guide. Washington, D.C.: U.S. Federal Aviation Administration, 1982.
54. Winter, P. M. Aviator's cancer. Lancet 2:285, 1968. (letter)
55. Wolf, C. R. Aerotitis in air travel. Calif. Med. 117(5):10–12, 1972.

7

Desirability and Feasibility of Additional Data Collection

The available empirical evidence is of insufficient quality and quantity for a scientific evaluation of airliner cabin air or of the probable health effects of short or long exposure to it. The Committee believes that this situation should be rectified and that data should be collected on the quality of airliner cabin air and on its health effects on passengers and crew.

There is a lack of definitive data showing relationships between airliner cabin air quality and health problems, except symptoms (chest pain, difficulty of breathing, and persistent cough) presumably associated with high ozone concentrations. Data are lacking because adequate studies have not been performed.

Several previous chapters have addressed the information relevant to the assessment of potential health risks associated with airliner cabin air: Chapter 2 described the systems for controlling cabin air, Chapter 5 described contaminants and special conditions of cabin air and the health effects usually associated with them, and Chapter 6 reviewed available evidence on the manifestation of health effects in crew or passengers. This chapter addresses the desirability and feasibility of collecting data that could be used to evaluate the quality of airliner cabin air and health effects associated with it. A start at modeling the most important factors that affect pollutant concentrations and flows on aircraft may be found in Appendix A. However, further model development and verification require a variety of additional data.

GENERAL CONCEPTS AND APPROACHES

The Committee has identified several potential sources of environmental quality problems on aircraft, including reduced air pressure, low humidity, ozone, cosmic radiation, and air contaminants, such as microbial aerosols. Although these factors are found in other environments as well, their combination in the aircraft cabin constitutes an environment whose uniqueness makes it difficult to draw valid conclusions on the basis of data on other environments. Although both the limited data available and calculations based on aircraft design and engineering information suggest that cabin air is probably no worse than air in many other confined environments, such a conclusion must remain speculative until valid measurements are made in the airliner cabin environment. The Committee believes that it is of paramount importance to measure characteristics of cabin air, to determine how they compare with conditions that cause problems in other environments.

Simply measuring the contaminants and other relevant variables of the airliner cabin does not address the question of the likely health effects of short or long exposure to that environment. The evaluation of the health effects of exposure requires the collection and interpretation of data very different from those on exposure. Furthermore, because it is difficult to detect and measure such effects, it is generally necessary to rely on measures that indicate or are related to the health effects of concern. The collection of data must be discussed with respect to four interrelated issues: potential causes of diminution in air quality, potential health effects of diminished air quality, actual examples of such effects, and surrogate measures of the effects where direct measurement is not possible. Extensive data on the operation and maintenance of aircraft have already been collected. The existing mechanisms of data collection should be examined to determine whether they can be used to satisfy these new needs.

Several parts of the federal regulations governing commercial air carriers[2] specify records and reports that commercial operators and air carriers must keep and submit to FAA. They include mechanical reliability reports describing the occurrence or detection of each failure, malfunction, or defect that endangers the safe operation of an aircraft.[4] Each certificate holder must submit a report covering each 24-h period to the FAA maintenance inspector assigned to its operations. In addition, summary reports on mechanical interruptions, alterations, and repairs must be submitted regularly,[1 3] and an airworthiness log kept on each aircraft must record all work performed on it, including maintenance, preventive maintenance, and alterations. Given the large numbers of aircraft in the fleets and the numbers of flights each day, these requirements generate a tremendous amount of data that provide a precise record that can be examined when accidents occur.

These data are entered into computerized storage and retrieval systems like the FAA Accident/Incident Data System (AIDS) and Service Difficulty Reports. However, such unfocused collection of information about almost anything that happens to each aircraft is difficult to use. Unless the data are classified according to relevant categories, it is very difficult to retrieve them in a way that is useful to answer the question under consideration. The FAA data collection and storage systems are oriented toward mechanical interruptions and accidents or incidents involving potential damage or injury, and the Committee has found the vast data collected by FAA to be of little use in assessing the quality of air in airliner cabins or the potential health consequences of exposure to it. The Committee suggests that consideration be given to adapting this data collection system to include collection of data relevant to the assessment of cabin air quality.

The potential health effects of cabin air considered by the Committee to be of greatest concern are reproductive effects, chronic pulmonary disease, chronic heart disease, cancer (including leukemia), and infectious disease. These effects are often hard to detect, measure, and attribute to specific causes. The numerous reasons include the lack of baseline observations on most persons who fly, the lack of equivalent groups with which to compare them, difficulties of measuring individual exposures, ethical constraints on and practical infeasibility of experimentation with various characteristics of cabin air, imprecision of signs and symptoms of acute effects (such as chest tightness), and the rarity of most effects of concern.

The Committee has identified several measures that are related to the health effects of concern, including reproductive function (e.g., abortion and birth-defect rates), pulmonary function (e.g., chronic obstructive pulmonary disease and disability), myocardial-infarction rates, use of onboard medical kits, and concentrations of specific contaminants (ozone, cosmic radiation, carbon monoxide, respirable suspended particles, and microorganisms). However, none of these measures has a one-to-one relationship with any of the health effects of concern, and most of the effects have several sources. Furthermore, data collected on health effects in airliner passengers or cabin crew will be extremely difficult to interpret, because of the difficulty of determining appropriate control groups. We know that the socioeconomic profile of the typical airline passenger is different from that of the general public, so we cannot be certain that the health effects observed in airline passengers are different from those in nonflyers, until they are compared with those in a similar group of nonflyers.

Despite these difficulties, the Committee concludes that appropriate data collection is not only possible, but highly desirable. The following sections describe the Committee's recommendations for research on airliner cabin air quality, the health effects of exposure to the cabin environment, and other topics.

MEASURES OF AIRLINER CABIN AIR QUALITY

The principal air quality problems on aircraft involve tobacco smoke, ozone, cosmic radiation, humidity, and microbial aerosols. Because ventilation rate and cabin pressure are the controlling factors for cabin air quality, actual ventilation rates should be measured under routine flight conditions in all types of commercial aircraft. The factors that influence pollutant concentrations and distribution within the cabin should be carefully considered, as well as the requirement of measuring concentrations over small spatial and temporal spans. If significant variations are found in an initial study, continual monitoring should be instituted.

Ozone is virtually the only source of degradation in air quality of which extensive measurements in aircraft have been reported. Exposure to ozone is regulated. Compliance can be achieved either through installation of filtration equipment (generally a catalytic converter), through the routing of flights so as to avoid areas of high ozone concentration (as detected by satellite), or through the choice of flight altitudes below those at which ozone is highly concentrated. The Committee feels that an evaluation of cabin air quality would be incomplete without a determination of the degree of compliance and the ozone concentrations to which passengers and cabin crew are exposed. The Committee accordingly recommends that FAA analyze cabin ozone concentrations. The analysis need not involve permanent monitoring, but should include sufficient data to provide a statistically representative sample of aircraft types, routes, and other factors relevant to the alternative ways of complying. Studies could be conducted in altitude chambers to determine whether ozone and the hypoxia induced by cabin pressurization to the equivalent of an 8,000-ft altitude are associated.

Exposure to cosmic radiation is a matter of concern. The Committee feels that FAA should periodically review flight routes and altitudes, to assess their implications for exposure to cosmic radiation. Regular representative sampling should be performed to estimate the exposure of the flying public. A special effort should be made to alert the medical profession to the hazards to groups that might be at increased risk, such as pregnant women and patients receiving particular medical therapy. Those who live at high altitudes should perhaps avoid further chronic exposure to cosmic rays in high-altitude flights. But such decisions require more reliable data than are available on the effects of chronic exposure to cosmic rays on the long-term incidence of neoplastic disease. Because routes change, FAA should measure exposure to cosmic rays on a representative sample of current flights.

The Committee strongly recommends that, so long as smoking is permitted in airplanes, the Congress mandate a program to monitor onboard carbon monoxide and respirable suspended particles. The Committee believes that, except for emergency situations involving fire, the most pervasive threat to airliner cabin air quality is cigarette smoke. Carbon monoxide and respirable suspended particles are two components of environmental tobacco smoke that are relatively easily measured, but the only empirical data have been collected on an ad hoc and nonrepresentative basis. There is a deficiency of information regarding hypoxia, which might result from synergism between altitude effects (decreased partial pressure of oxygen) and formation of carboxyhemoglobin (due to increased molar concentration of carbon monoxide). Studies are beginning to evaluate this interaction, but at higher ambient carbon monoxide concentrations than reportedly occur in the aircraft cabin. Patients with cardiorespiratory problems might be at greater risk, as might cabin attendants who must work and rest in these conditions.

Many people believe that one is more likely to catch cold or contract a respiratory infection in an airplane than in most other common environments, but no evidence has been produced to establish this. In view of the degree of expressed concern about microbial contamination in aircraft and the possibility that serious acute health effects could result from such contamination, it is important to collect baseline data on background concentrations of microbial aerosols during normal flight conditions. It is also important to collect data on microbial aerosols in aircraft with known emission sources and under conditions of decreased ventilation. The Congress should authorize and appropriate funds for studies to measure volumetrically bioaerosol concentrations and associated variables in aircraft in flight—such as temperature, relative humidity, ventilation rate, filtration modes, and number of passengers on board—and bioaerosol concentrations in intake air in aircraft on the ground.

The purpose of gathering data on the various potential contaminants of airliner cabin air is to compare the concentrations measured with those believed to cause health problems in other environments. Even though the combination of environmental conditions found on aircraft is unique, such comparisons can identify possible problems, which can then be examined in greater detail.

MEASURES OF HEALTH EFFECTS

The previous section identified several potential contaminants of airliner cabin air on which the Committee recommends collection of additional data. As pointed out earlier in this chapter, data on the potential health effects of these contaminants in the airliner environment must also be collected, but they must be collected and interpreted in ways that differ considerably from those for data on the contaminants.

The Committee attempted to identify measures for each of the health effects of concern: reproductive effects, chronic obstructive pulmonary disease, chronic heart disease, cancer, and infectious disease. However, direct measurement of these health effects is often not possible; therefore, collection of data on a series of suggestive measures is recommended.

Appropriately designed studies of selected health effects among crew members would be useful and ought to be performed, but finding valid comparison groups will be more difficult than in other industrial epidemiologic studies. For example, comparing disease rates of male employees in a particular factory with rates in the general population usually shows the workers to be healthier, because the total population includes all sick people. It might be better to compare the workers in one factory with those in another. But it is not possible to determine from the data on health alone which group of workers is exposed to the greater risk. That requires accompanying measures of exposure as well. Data on health effects of airliner cabin air in passengers pose even more problems, because relatively little is known about the characteristics of the flying public and it is not clear how to identify an equivalent group of people who do not fly. Even though the relevant characteristics of cabin crews are much better known, it is still difficult to find a group of nonflyers or infrequent flyers with whom appropriate comparisons can be made.

The Committee feels that, given the nature of the exposures and resulting health effects and the special occupational setting, it is unrealistic to expect that feasible epidemiologic studies will be able to determine conclusively the health hazards associated with exposure to airliner cabin air. Nevertheless, even though such studies cannot prove the degree of hazard associated with such exposure, they can produce data that are suggestive and that identify potential problems for further analysis.

The Committee recommends studies to examine rates of spontaneous abortion and birth defects among cabin crew members. Cabin crew members are subject to longer exposure than the flying public in general, and in examining reproductive effects it is not necessary to wait many years for chronic effects to emerge. In addition, reproductive effects are often sensitive indicators of other effects that are more difficult to measure. The only way to determine with accuracy whether the observed reproductive effects were due to exposure during flight, as opposed to exposure in the home or exposure to other personal variables, would be to assign new employees at random to cabin crews, as opposed to, say, work at ticket counters. The rates exhibited over time by the two groups would then be directly compared, to assess the reproductive hazards of exposure during flight. Such random assignment of employees is not practical. In lieu of it, comparisons would need to be made with several groups of similar ages, places of residence, family status, and other characteristics. Even then, the results could be considered only suggestive, and more detailed examinations would be required if problems were revealed. Care would need to be exercised to ensure that the groups examined were large enough to permit statistically significant analyses, and it could prove extremely difficult to find groups that include enough people with appropriate characteristics. In addition, careful measurements of exposure (or appropriate surrogates) should be made. Despite the difficulties in interpreting results, the Committee recommends that a feasibility study be undertaken to determine whether these conditions can be met.

The Committee feels that it is important to test pulmonary function among crew members and perhaps among selected passengers. In particular, chronic obstructive pulmonary disorders and pulmonary disability should be identified. The Committee feels that both flow-loop volume tests and forced expiratory volume (FEV) tests should be used. Flow-loop tests require more sophisticated computer equipment and are less susceptible to intentional or unintentional manipulation by subject or observer. However, FEV tests have been used successfully in many epidemiologic studies and would permit comparison with results under other conditions. Flight attendants have consistently reported respiratory effects, probably because their activity is greater than that of passengers. Studies in which subjects are exposed to ozone and carbon monoxide clearly indicate that the combination of exposure and increased exercise results in increased effects on cardiopulmonary function. The Committee feels that data concerning effects on pulmonary function would be vital in evaluating the health effects of airliner cabin air and recommends that appropriate before-and-after testing be undertaken.

It is difficult to determine an appropriate approach to the gathering of data on the incidence of myocardial infarction associated with air travel. The onset of myocardial infarction might be a response more to the stress of flying than to exposure to cabin air. Furthermore, the period at hazard may extend from before boarding to after deplaning. Most large airports have emergency medical facilities of some sort, so it might be possible to gather data on the incidence of myocardial infarction in or near airports and compare that incidence with the incidence during flight. The Committee feels that such a study is important enough to require a feasibility study to determine whether accurate data in sufficient quantity could be collected.

Measures for cancer are impractical, because of the long period of latency between exposure and onset. Although shorter, the incubation period for most infectious diseases precludes development of measures of them as well. However, from the standpoint of occupational health, it is entirely feasible and important to undertake a prospective monitoring of exposures and eventual mortality based on the National Death Index.

OTHER SUBJECTS

On January 9, 1986, FAA published a final rule requiring an approved medical kit to be carried on all passenger flights, training to familiarize crew members with the kit, and the reporting to FAA of each medical emergency during flight that results in use of the kit for the first 24 mo after the effective date of the rule.[5] The Committee recommends that FAA—in conjunction with physicians, statisticians, and epidemiologists—establish a clear protocol for reporting data on the use of emergency medical kits. If collection procedures are properly designed, the resulting data can be analyzed to identify the pattern of medical incidents during flight and to compare these patterns with the incidence of emergencies in other settings.

The Committee also feels that it would be advisable to monitor scientific literature relevant to various aspects of airliner cabin air quality or its health effects. Available computer-based bibliographic databases, such as MEDLINE, could be easily and inexpensively searched regularly to identify new scientific developments relevant to the topics addressed in this report.

REFERENCES

1. Airworthiness release or aircraft log entry. Code of Federal Regulations, Title 14, Pt. 121.709. Washington, D.C.: U.S. Government Printing Office, 1985.
2. Certification and operations: Domestic, flag and supplemental air carriers and commercial operators of large aircraft. Code of Federal Regulations, Title 14, Pt. 121. Washington, D.C.: U.S. Government Printing Office , 1985.
3. Mechanical interruption summary report. Code of Federal Regulations, Title 14, Pt. 121.705. Washington, D.C.: U.S. Government Printing Office, 1985.
4. Mechanical reliability reports. Code of Federal Regulations, Title 14, Pt. 121.703. Washington, D.C.: U.S. Government Printing Office, 1985.
5. U.S. Federal Aviation Administration. Emergency medical equipment: Final rule. Federal Register 51(Jan. 9):1218–1223, 1986.

Appendix A

A Computer Model for Assessing Airliner Cabin Air Quality

Full understanding of cabin air quality requires, among other things, the monitoring of various pollutant concentrations. That is difficult and costly, because so many different pollutants require different monitoring devices and protocols. It would not be cost-effective to study all possible pollutants, although similarities in the sources and sinks of some pollutants would eliminate the necessity of monitoring all of them, and some pollutants are likely to be present in such low concentrations as to be unmeasurable and unimportant with respect to health or welfare.

The prohibitive cost of an extensive monitoring program suggests that we look for a different approach to assessing cabin air quality. A model of cabin air quality could serve adequately as an investigative tool. An accurate, validated model could be used to pinpoint potential problems and to study the sensitivity of pollutant concentrations to various control measures. The costs associated with control methods can be estimated with a separate model. With the results of modeling pointing the way, the attack on the problem could be more focused.

CONCEPTUAL DEVELOPMENT

The model must account for the important aspects of cabin air quality. It must be flexible enough to be used for various types of pollutants with different source profiles, temporal patterns, and health implications. It must be accessible to persons unfamiliar with mathematical or computer modeling. In fact, the details of the model need not be known to the user; only the outcome need be analyzed.

The question to be answered is simply stated: Given a few external characteristics, estimate the concentration of a pollutant in the cabin. Several physical characteristics are available to the modeler. Aircraft volumes and air-movement systems are well defined. Ventilation of the cabin is an energy-using process, the engineering designs are well optimized, and data are available. Information on air recirculation and filtering is also available, as is information on the source strengths of some of the pollutants, such as carbon dioxide and water vapor from humans, tobacco smoke, and ozone. Less is known about others, such as volatile organic compounds emitted from materials, insecticides, or cleaning agents.

Other input data for the model are not readily available. These include information on air-mass movements between compartments in the cabin, rates of loss of reactive chemicals, and chemical deposition rates. These qualities can be estimated, but an effective model must include the ability to perform sensitivity analyses for them. Ideally, it should be possible to perform sensitivity analyses as the need arises for all quantities on which little information is available or for which design specifications are not met.

Once the potential input data are known, selection of a model type can begin. The most appropriate type of model for this application should be based on the general mass-balance approach. All mechanisms for production and loss of the pollutant are accounted for properly, and the change in concentration per unit time is the difference between the two:

(1)

$dC/dt = P-LC$,

where C = concentration of pollutant,

t = time,

P = rate of production of pollutant, and

L = rate of loss of pollutant.

At equilibrium, $dC/dt = 0$—production rate equals loss rate. Solving for the concentration gives:

(2)

$C = P/L$.

Note that the ratio 1/L is a measure of the lifetime of the exponential approach to equilibrium. Small rates of loss imply a slow approach to equilibrium; large rates of loss suggest that equilibrium will be established rapidly and will prevail.

Figure A-1 is a schematic of a single component of a multibox model of an aircraft. The model consists of essentially separate boxes, each containing its own production and loss mechanisms. Production mechanisms include pollutant presence in circulating air (R_i) and local source (S_i). Potential production mechanisms from reactive chemistry can be added, although they are probably unimportant. Loss mechanisms include leakage (L_i), main recirculating flow (F_i), and first-order losses from deposition or other processes (K_i). An assumed local equilibrium is established in each. A degree of communication is established between adjacent boxes only because of the presence of small forward and backward airflow terms (f_i and b_i). These terms act as additional loss mechanisms for the box in question, whereas terms from adjacent boxes act as production mechanisms.

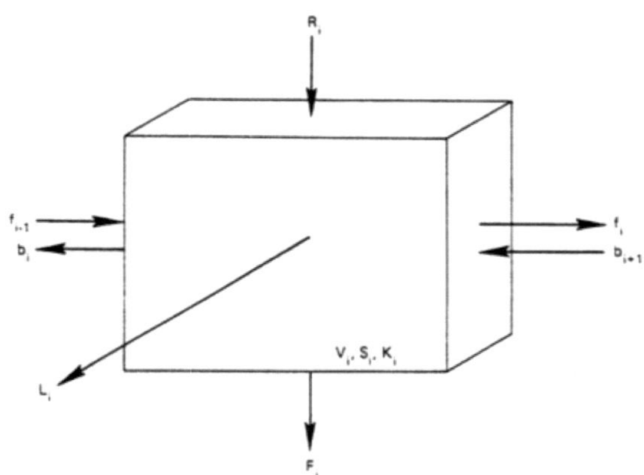

FIGURE A-1
Schematic of single component of multibox model of aircraft cabin air quality. See text for explanation of symbols.

The cabin itself is coupled to another system within the aircraft, the air cleaning system. Figure A-2 is a schematic of the aircraft as a whole. Note that the cabin can be considered to be a single compartment (box 0), with polluted air leaving the cabin (F) and entering the air cleaning system. There a portion of the polluted air (E) is exhausted, and the remainder is filtered, mixed with ambient makeup air (m), and returned to the cabin (R) as the supply air. Conservation of mass requires that F + L + E = m + R, where L is the amount leaked from the cabin to the atmosphere.

More complicated systems require more complicated analysis. If all pollutants are generated in one place, but other places are of interest, more boxes are needed to describe the system. As the system becomes more

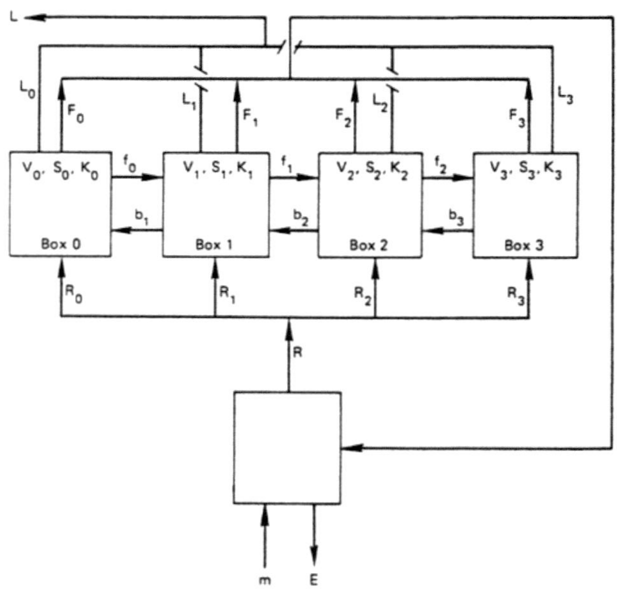

FIGURE A-2
Schematic of aircraft with air cleaning system. See text for explanation of symbols.

complex, more information is needed for the model. Data on exchange of air from one box to another must be obtained. Analysis of the results also becomes more complex. Figure A-3 illustrates the detailed physical model schematically. In this case, four of the detailed boxes are coupled within the cabin. Note that no forward flow (f_i) is allowed out of the foremost compartment, nor is any backward flow (b_i) allowed out of the rearmost compartment. This physical model is very general. Each compartment can have any volume (V_i) deemed appropriate. Preferential flow can be effected by manipulating the relative magnitudes of f_i and b_i. Differential source strengths can also be implemented. Additionally, control strategies and their economic impacts can be investigated.

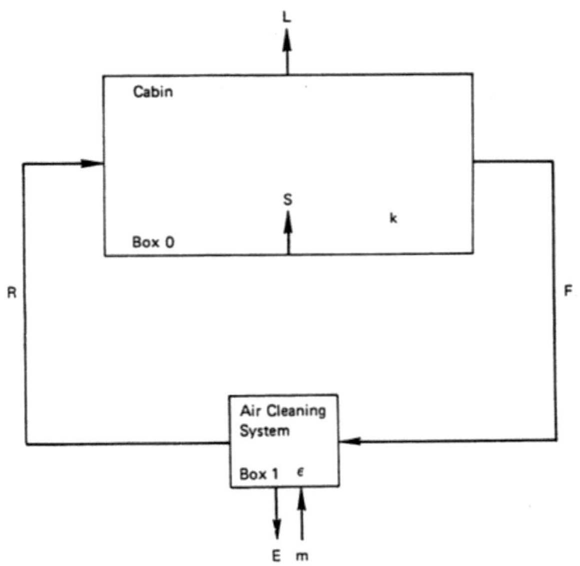

FIGURE A-3
Schematic of coupled components for multibox model of aircraft cabin air quality. See text for explanation of symbols.

MATHEMATICAL DEVELOPMENT

The determination of the concentrations in each of the compartments in the model described above requires the simultaneous solution of coupled, first-order linear differential equations obtained from Equation 1. At equilibrium, the solution is easily cast into the form of a matrix equation. Because of the nature of the physical model—i.e., interaction of adjacent boxes only—the mathematical form is tractable. Solutions can be obtained quickly and accurately for a large number of interacting boxes.

To describe the system, start with an expansion of Equation 1 for the 1th box.

$$\frac{dC_i}{dt} = \frac{S_i}{V_i} + \frac{R_i}{V_i} + \frac{f_{i-1}}{V_i} - \frac{1}{V_i}(L_i + f_i + b_i + F_i) + K_i \qquad (3)$$

The restrictions on fi and bi apply. At equilibrium, all dCi/dt vanish, and the matrix equation becomes (for a four-compartment case):

$$\begin{bmatrix} \frac{1}{V_0}(L_0 + f_0 + F_0) + K_0 & -\frac{b_1}{V_0} & 0 & 0 \\ -\frac{f_0}{V_1} & \frac{1}{V_1}(L_1 + f_1 + F_1) + K_1 & -\frac{b_2}{V_1} & 0 \\ 0 & -\frac{f_1}{V_2} & \frac{1}{V_2}(L_2 + f_2 + F_2) + K_2 & -\frac{b_3}{V_2} \\ 0 & 0 & -\frac{f_2}{V_3} & \frac{1}{V_3}(L_3 + f_3 + F_3) + K_3 \end{bmatrix} \begin{bmatrix} C_0 \\ C_1 \\ C_2 \\ C_3 \end{bmatrix}$$

$$= \begin{bmatrix} \frac{1}{V_0}(S_0 + R_0 C_r) \\ \frac{1}{V_1}(S_1 + R_1 C_r) \\ \frac{1}{V_2}(S_2 + R_2 C_r) \\ \frac{1}{V_3}(S_3 + R_3 C_r) \end{bmatrix} \qquad (4)$$

where C_r represents the concentration of pollutant in the recirculated air. This system has a tridiagonal form and can be solved efficiently with LU factorization. The coupled cabin and air cleaning system is solved first, with an explicit solution of the two-by-two form.

OPERATING PROCEDURES

The cabin air quality simulation model Cabinair is designed to be user-friendly and self-documenting. The operator specifies whole aircraft parameters as listed in Table A-1. Any of these parameters can be changed through commands. It is important, however, that consistency checks be made to ensure mass balance, etc. A warning is displayed when, for example, total flow in exceeds total flow out. Table A-1 lists a standard set of parameters programed as default values. These are appropriate for an L-1011 with four compartments and tobacco smoke as the pollutant of interest.

TABLE A-1 Parameters for Whole Aircraft with L-1011 Four-Zone Parameterization

Parameter	Value
Volume	450.0 m^3
Recirculation	150.0 m^3/min
Leak rate	10.0 m^3/min
Net flow rate	140.0 m^3/min
Deposition	0.0033/min
Source rate	83.3300 mg/min
Exhaust flow	140.0 m^3/min
Makeup flow	150.0 m^3/min
Outdoor concentration	0.01000 mg/m3
Number of boxes	4

SIMULATING AIRLINER AIR QUALITY

The Cabinair model was used to simulate the steady-state concentrations of environmental tobacco smoke, carbon dioxide, and water vapor in multiple zones of three aircraft: B-727-200, B-767-200, and MD-80. Flow parameters were developed from the technical ventilation specifications of the aircraft. Figures A-4, A-5, and A-6 show the outside-air supply, recirculation, and controlled and uncontrolled leakage for these three aircraft.

The B-727-200 (Figure A-4) has a straightforward once-through ventilation system. The ECUs deliver 240 cfm, 350 cfm, and 2,235 cfm to the cockpit, first-class section, and coach section, respectively. The outside air delivered to the passenger sections (first class and coach) is assumed to be delivered uniformly over the entire length of the cabin. Air is discharged through both controlled and uncontrolled vents. The aft exhaust valve is used to control pressure and discharges 883 cfm. Avionics, cargo, lavatory, and galley vents (forward and aft) discharge a total of 892 cfm. There is leakage of 1,050 cfm.

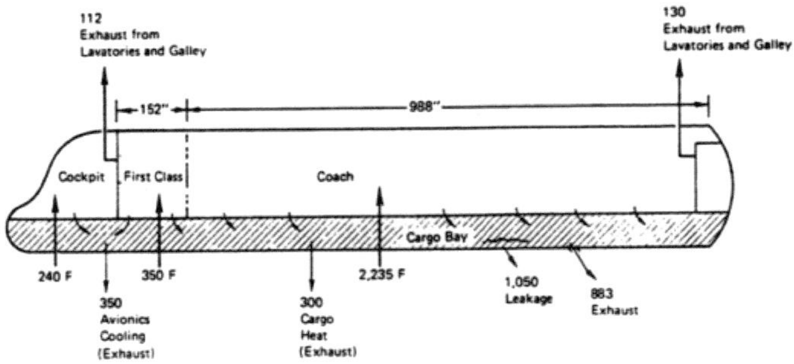

FIGURE A-4
B-727-200 cabin airflow distribution, cfm. All outside air. F, outside air. Uniform supply in cabin. Exhaust uniform at floor level. Leakage assumed uniform at 1,050 cfm. Arrows show direction of airflow. Based on information from Boeing (personal communication, 1985) and Lorengo and Porter.[3]

The B-767-200 (Figure A-5) has a more complex ventilation system. The 2,388 cfm from the ECUs is mixed with 2,388 cfm of filtered recirculation air from the forward cabin and delivered to the cockpit and to the overhead air vents in the cabin. The overboard discharge manifold draws air from lavatories, galleys, and the aft avionics compartments, which then mixes with floor-level cabin exhaust and is discharged overboard.

The MD-80 ventilation system (Figure A-6) is different from either of the other two. The cockpit is supplied only with outside air. Recirculation air is drawn from along the floor of both first-class and coach sections of the passenger cabin and is mixed with outside air. This mixed air is then delivered to the first-class and coach sections.

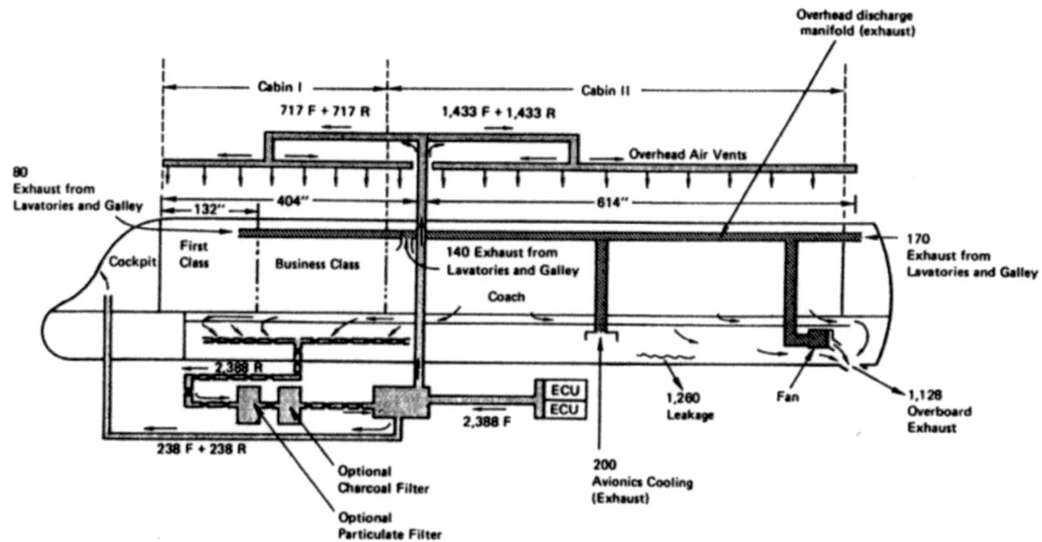

FIGURE A-5
B-767-200 cabin airflow distribution, cfm. ECU, environmental control unit. F, outside air. R, recirculated air. Arrows show direction of airflow. Leakage assumed uniform at 1,260 cfm. Based on information from Boeing (personal communication, 1985) and Lorengo and Porter.[3]

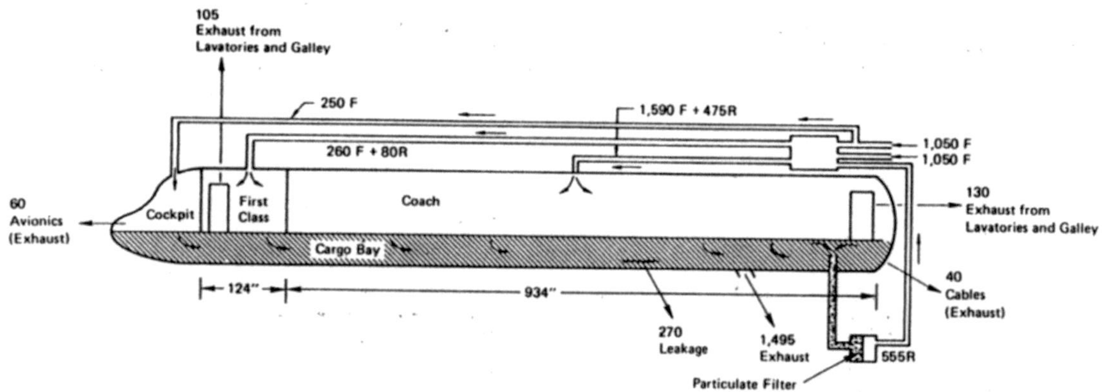

FIGURE A-6
MD-80 cabin airflow distribution, cfm. F, outside air. R, recirculated air. Leakage assumed uniform at 270 cfm. Arrows show direction of airflow. Based on information from McDonnell Douglas (personal communication, 1985) and Lorengo and Porter.[3]

With these data and standard configuration diagrams available from Trans World Airlines (Figures A-7, A-8, and A-9), a volume-weighted partitioning of the flows was made. The volume of a given zone was assumed to be directly proportional to the linear dimension of the zone as a fraction of the total length of the aircraft. Generally, after the initial partitioning, flow imbalances remained. These imbalances were eliminated by allowing forward or backward flow to or from adjacent zones to compensate for an excess or deficiency of air movement. Flows were thus balanced to within approximately 1 m^3/min over the entire aircraft.

The source strengths used in the simulations were as follows. For respirable particles, cigarette smoke is the primary source. An active smoker produces

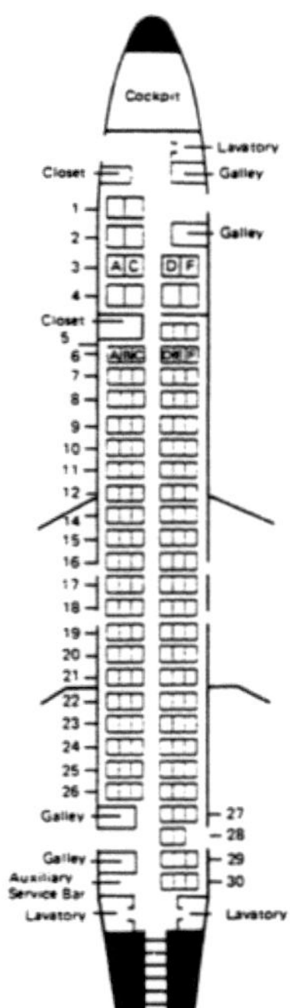

Zone	Rows	Seats	Crew
Flight Deck	—	—	3
First Class Lavatory and Galley	—	—	—
First Class Non-Smoking	1-3	8	1
First Class Smoking	4	4	—
Coach Non-Smoking	5-19	81	2
Coach Transition	20-21	12	—
Coach Smoking	22-30	41	1
Coach Lavatory and Galley	—	—	—

FIGURE A-7
Standard B-767-200 interior arrangement. Numbers of rows allotted for smoking can be increased or reduced according to demand for nonsmoking seats. Figure reprinted with permission from Trans World Airlines.[4]

respirable particles at approximately 3 mg/min. On the average, a smoker smokes 2 cigarettes/h and takes 10 min/cigarette, thus smoking one-third of the time. The scenarios investigated include an average state in which one-third of the smokers are smoking or every smoker is smoking at one-third the maximal rate. At a maximum, all smokers are smoking simultaneously. For carbon dioxide, a source strength of 0.5 L/min per person is used (a source strength of 0.5 L/min per person is used to approximate the proportions of active crewmembers and sedentary passengers).[2] ASHRAE uses 0.3 L/min.[1] The figure of 0.5 L/min is equivalent to 760 mg/min per person. A sedentary person, such as a passenger,

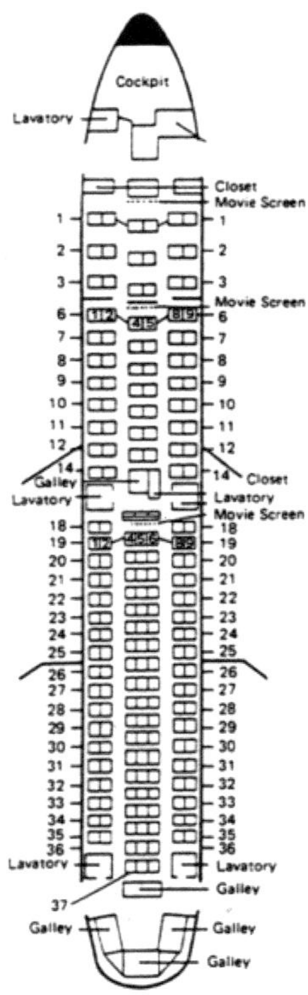

Zone	Rows	Seats	Crew
Flight Deck	—	—	2
First Class Lavatory and Galley	—	—	—
First Class Non-Smoking	1-2	12	1
First Class Smoking	3	6	1
Business Non-Smoking	6-10	30	1
Business Smoking	11-14	16	1
Business Lavatory and Galley	—	—	—
Coach Non-Smoking	18-27	67	1
Coach Transition	28-29	18	—
Coach Smoking	30-37	48	1
Coach Lavatory and Galley	—	—	—

FIGURE A-8
Standard B-767-200 interior arrangement. Numbers of rows allotted for smoking can be increased or reduced according to demand for nonsmoking seats. Figure reprinted with permission from Trans World Airlines.[4]

produces water vapor at 700 mg/min, whereas an active person, such as a crew member, produces 2,000 mg/min.

Ambient concentrations of particles, carbon dioxide, and water are 0.010 $\mu g/m^3$, 330 ppm, and 1.5 g/kg of air, respectively. Changing these values (by filtration) will alter the results only slightly for respirable particles, but might have larger effects for carbon dioxide and water vapor.

Data in Tables A-1 through A-3 are for aircraft in the standard configuration, including normal recirculation and full occupancy with all packs running. Table A-4 presents data on the MD-80 aircraft, assuming, for comparative purposes, no recirculation. Table A-5 presents data on the B-767-200 aircraft with no recirculation, and Table A-6 presents data on the B-767-200 aircraft, assuming standard operating conditions, but only 60% occupancy.

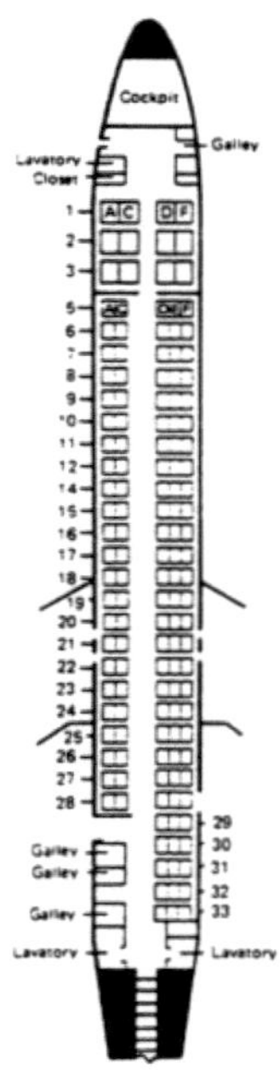

Zone	Rows	Seats	Crew
Flight Deck	—	—	2
First Class Lavatory and Galley	—	—	—
First Class Non-Smoking	1-2	8	1
First Class Smoking	3	4	—
Coach Non-Smoking	5-22	85	2
Coach Transition	23-24	10	—
Coach Smoking	25-33	35	1
Coach Lavatory and Galley	—	—	—

FIGURE A-9
Standard MD-80 interior arrangement. Numbers of rows allotted for smoking can be increased or reduced according to demand for non-smoking seats. Figure reprinted with permission from Trans World Airlines.[4]

Generally, when air is recirculated, the concentrations of pollutants increase. As occupancy decreases, the concentrations of pollutants decrease. Although it is not exact, one can approximate both these phenomena as linear; i.e., 50% recirculation will result in doubling the pollutant concentrations, and 50% occupancy will halve the concentrations.

TABLE A-2 Calculated Concentrations of Various Pollutants on Simulated B-727-200 Aircraft

	Environmental Tobacco Smoke, mg/m^3			
Zone[a]	Average[b]	Maximum[c]	CO_2, ppm	Relative Humidity (Water Vapor), %
Cockpit	0.010	0.010	517	7.8
First-class lavatory and galley	0.010	0.010	435	5.5
First-class nonsmoking	0.010	0.010	919	8.7
First-class smoking	1.302	3.886	1,178	10.8
Coach nonsmoking	0.058	0.154	1,284	10.9
Coach transition	0.018	0.034	1,373	11.6
Coach smoking	2.243	6.708	1,367	11.6
Coach lavatory and galley	0.299	0.876	484	4.4
Whole aircraft[d]	0.560	1.661	1,139	10.1
Volume averaged[e]	0.570	1.691	1,154	10.1
Supply air	0.010	0.010	330	3.7

[a] Zones are examples of standard configuration zones; 100% occupancy assumed; no recirculation. Supply air concentration is ambient concentration. CO_2 and water vapor concentrations assume temperature of 20°C.

[b] One-third of cigarette smokers smoking at any time (2 cigarettes/h).

[c] All cigarette smokers on plane smoking at same time.

[d] Average concentration derived from arithmetic average of zonal concentrations.

[e] Derived from zonal concentrations weighted by volume.

TABLE A-3 Calculated Concentrations of Various Pollutants on Simulated B-767-200 Aircraft

	Environmental Tobacco Smoke, mg/m^3			
Zone[a]	Average[b]	Maximum[c]	CO_2, ppm	Relative Humidity (Water Vapor), %
Cockpit	0.297	0.872	770	7.2
First-class lavatory and galley	0.295	0.865	770	6.7
First-class nonsmoking	0.293	0.860	1,240	10.3
First-class smoking	1.196	3.569	1,469	12.0
Business-class nonsmoking	0.471	1.395	1,535	11.5
Business-class smoking	1.998	5.976	1,590	11.8
Business-class lavatory and galley	0.314	0.923	1,140	8.2
Coach nonsmoking	0.293	0.861	1,483	11.4
Coach transition	0.293	0.861	1,773	13.7
Coach smoking	2.380	7.122	1,662	12.8
Coach lavatory and galley	0.660	1.961	1,610	6.6
Whole aircraft[d]	0.798	2.375	1,354	10.5
Volume averaged[e]	0.827	2.461	1,389	10.7
Supply air	0.300	0.881	707	6.2

[a] Zones are examples of standard configuration zones; 100% occupancy assumed; 50% of return air recirculated. CO_2 and water vapor concentrations assume temperature of 20°C.

[b] One-third of cigarette smokers smoking at any time (2 cigarettes/h).

[c] All cigarette smokers on plane smoking at same time.

[d] Average concentration derived from arithmetic average of zonal concentrations.

[e] Derived from zonal concentrations weighted by volume.

TABLE A-4 Calculated Concentrations of Various Pollutants on Simulated MD-80 Aircraft

	Environmental Tobacco Smoke, mg/m^3			
Zone[a]	Average[b]	Maximum[c]	CO_2, ppm	Relative Humidity (Water Vapor), %
Cockpit	0.126	1.214	638	7.6
First-class lavatory and galley	0.125	0.784	599	6.3
First-class nonsmoking	0.125	0.577	965	9.2
First-class smoking	0.688	2.209	867	10.1
Coach nonsmoking	0.206	0.634	1,522	12.6
Coach transition	0.638	1.912	1,585	12.7
Coach smoking	2.237	6.710	1,452	11.9
Coach lavatory and galley	0.124	0.370	540	4.5
Whole aircraft[d]	0.631	1.968	1,329	11.2
Volume averaged[e]	0.593	1.850	1,270	10.8
Supply air	0.127	0.380	519	5.1

[a] Zones are examples of standard configuration zones; 100% occupancy assumed; 21% of return air recirculated. CO_2 and water vapor concentrations assume temperature of 20°C.

[b] One-third of cigarette smokers smoking at any time (2 cigarettes/h).

[c] All cigarette smokers on plane smoking at same time.

[d] Average concentration derived from arithmetic average of zonal concentrations.

[e] Derived from zonal concentrations weighted by volume.

TABLE A-5 Calculated Concentrations of Various Pollutants on Simulated B-767-200 Aircraft with No Recirculation

	Environmental Tobacco Smoke, mg/m^3			
Zone[a]	Average[b]	Maximum[c]	CO_2, ppm	Relative Humidity (Water Vapor), %
Cockpit	0.010	0.010	393	4.9
First-class lavatory and galley	0.010	0.010	393	4.6
First-class nonsmoking	0.010	0.010	863	8.3
First-class smoking	0.914	2.721	1,091	10.1
Business-class nonsmoking	0.190	0.549	1,157	9.6
Business-class smoking	1.717	5.131	1,212	9.9
Business-class lavatory and galley	0.315	0.075	762	6.3
Coach nonsmoking	0.010	0.010	1,105	9.4
Coach transition	0.010	0.010	1,394	11.7
Coach smoking	2.097	6.271	1,283	10.8
Coach lavatory and galley	0.376	1.110	518	4.6
Whole aircraft[d]	0.515	1.525	976	8.5
Volume averaged[e]	0.544	1.611	1,011	8.7
Supply air	0.010	0.010	330	3.7

[a] Zones are examples of standard configuration zones; 100% occupancy assumed. CO_2 and water vapor concentrations assume temperature of 20°C.

[b] One-third of cigarette smokers smoking at any time (2 cigarettes/h).

[c] All cigarette smokers on plane smoking at same time.

[d] Average concentration derived from arithmetic average of zonal concentrations.

[e] Derived from zonal concentrations weighted by volume.

TABLE A-6 Calculated Concentrations of Various Pollutants on Simulated B-767-200 Aircraft with 60% Occupancy and 50% Recirculation

	Environmental Tobacco Smoke, mg/m^3			
Zone[a]	Average[b]	Maximum[c]	CO_2, ppm	Relative Humidity (Water Vapor), %
Cockpit	0.182	0.527	521	5.5
First-class lavatory and galley	0.181	0.523	483	5.1
First-class nonsmoking	0.180	0.520	681	7.2
First-class smoking	0.722	2.145	777	8.3
Business-class nonsmoking	0.287	0.841	747	7.9
Business-class smoking	1.203	3.589	760	8.1
Business-class lavatory and galley	0.192	0.558	563	6.0
Coach nonsmoking	0.180	0.520	747	7.9
Coach transition	0.180	0.520	877	9.3
Coach smoking	1.432	4.277	827	8.8
Coach lavatory and galley	0.400	1.180	473	5.0
Whole aircraft[d]	0.483	1.525	693	7.5
Volume averaged[e]	0.500	1.429	706	7.5
Supply air	0.184	0.533	474	5.0

[a] Zones are examples of standard configuration zones. CO_2 and water vapor concentrations assume temperature of 20°C.

[b] One-third of cigarette smokers smoking at any time (2 cigarettes/h).

[c] All cigarette smokers on plane smoking at same time.

[d] Average concentration derived from arithmetic average of zonal concentrations.

[e] Derived from zonal concentrations weighted by volume.

REFERENCES

1. American Society of Heating, Refrigerating, and Air-Conditioning Engineers, Inc. ASHRAE Standard: Ventilation for Acceptable Air Quality. ASHRAE 62-1981. Atlanta, Ga.: American Society of Heating, Refrigerating, and Air-Conditioning Engineers, Inc., 1981.
2. Balvanz, J., S. C. Bowman, A. Fobelets, T. Lee, K. Papamichael, and R. Yoder. Examination of the Cabin Environment of Commercial Aircraft. Ames, Iowa: Iowa State University, 1982.
3. Lorengo, D. E., and A. Porter. Aircraft Ventilation Systems Study: Final Report. DTFA-03-84-CO-0084. Atlantic City, N.J.: U.S. Federal Aviation Administration Technical Center, 1985. (draft)
4. Trans World Airlines. TWA Aircraft Seating Guide. New York, N.Y.: Trans World Airlines, 1984.

Appendix B

Selected Material from the FAA Accident/ Incident Data System

This appendix presents data from the FAA Accident/Incident Data System (AIDS) on in-flight fires and explosions and ground fires (Table B-1), cabin smoke (Table B-2), and emergency descent and deployment of oxygen masks (Table B-3). The incidents summarized here constitute all those reported from the beginning of 1980 through November 1985. The only deaths reported were those associated with the Air Canada fire near Cincinnati in 1983, in which 23 people died.

The data appear essentially as they appear in the AIDS computer printout. Some explanation of codes and abbreviations used in the tables is in order. The date of an incident is presented as a six-digit number, in which the first two digits represent the year, the next two the month, and the final two the day of the month; for example, 800118 means January 18, 1980. The aircraft manufacturers' names need little explanation, but "Doug" stands for McDonnell Douglas, "CVAC" for Convair Aircraft Company, "Airbus" for the European manufacturer Airbus Industrie, "BAC" for British Aerospace Corp., "EMB" for the Brazilian Embraer, "StBros" for the Irish firm Short Bros., and "Swrngn" for Sweringen. "TOB" stands for total on board. In the "Damage" column, "N" means none, "S" means slight, "M" means moderate, and "D" means severe. Under "Flt/Type" (flight type), "APAX" means all passenger, "ACAR" means all cargo, "PXCG" means combined passenger and cargo, and "OTHER" covers all remaining slight categories, such as test and experimental flights.

TABLE B-1

In-flight Fires and Explosions and Ground Fires

Date	Make	Model	TOB	Damage	Flt/Type	Remarks
800118	Nihon	YS11A500	8	N	APAX	Right engine fire warning came on in flight. Discharged fire bottle and shut down engine.
800119	Boeing	B-747	3	S	ACAR	Pilot shut down #3 engine because of fire warning light. Restarted later in flight. Fire damage found.
800228	Doug	DC-8	108	M	APAX	Engine shutdown due to power loss and vibration. Tach generator housing had been ingested into the engine.
800319	Boeing	B-737-222	75	N	APAX	Loose landing light switch. Terminal caused minor electrical fire.
800406	Boeing	B-747-228B	111	N	APAX	Flight returned due to fire warning light and engine shut down. Engine was damaged.
800514	Doug	JT3D	94	M	APAX	Engine failure resulted in numerous punctures in skin. Fire started due to ruptured fuel tank.
800614	Boeing	B-747-121	111	M	APAX	Engine shut down. Flight returned due to fire warning. Found chafed hydraulic line at engine started fire.

Date	Make	Model	TOB	Damage	Flt/Type	Remarks
800626	Doug	DC-9-31	3	M	APAX	Right engine compressor failed causing fire.
801008	Doug	DC-9-31	60	M	APAX	Passenger's lighter burst into flames. Flight attendant extinguished the fire with apple juice.
801205	Boeing	B-727-227	57	M	APAX	Several breakers opened and crew smelled smoke after takeoff. Returned.
801211	Lkheed	L-1011	108	M	APAX	Experienced engine fire during taxi. Suspect faulty generator.
810128	Doug	DC-10	83	M	APAX	Fire occurred in passenger carry on hand bag. Probably due to careless smoking.
810511	Boeing	B-727-231	7	M	APAX	Smoke and flames behind forward instrument panel. Found master warning card and wiring burned.
810617	Boeing	B-727-23	1	N	APAX	Baggage fire in flight. Found package with 24 volt battery pack which may have shorted. No aircraft damage.
810706	CVAC	CVAC-340	19	N	APAX	Fire warning/bell came on when #2 engine shutdown. Oil accumulations from small leaks ignited around augmentor.

810725	CVAC	STC240	2	M	ACAR	Found a fire in the left engine. Damage confined to engine compartment.
810814	Lkheed	L-1011	1	M	APAX	Electrical fire in service cart receptable. Put out via CO_2 and unplugged. Continued to destination.
810817	Boeing	B-747	427	N	APAX	Inflight fire in lavatory trash bin. Extinguished by cabin crew. No damage.
810828	Boeing	B-747-135	287	N	APAX	Cabin wall fire from exploding window light ballast.
810914	Boeing	B-747-151	208	M	APAX	Compressor stall and engine fire NR4 engine. Used extinguishers. Landed ok. NR2 bearing and NR4 gear box failure.
810914	Boeing	B-747	92	M	APAX	#3 engine fire after takeoff. Returned. Thrust reverser bleed valve broken. Burned wiring in stub wing.
811212	Boeing	B-727-223	32	N	APAX	Holding near gate, started APU, fire seen by cabin attendant, APU replaced. No fault found.
811212	Boeing	B-747	3	M	PXCG	On approach a small fire erupted in upper galley waste container. Put out by ice water. Fire bottle empty.

Date	Make	Model	TOB	Damage	Flt/Type	Remarks
820202	CVAC	STC204T	2	M	ACAR	Fire right engine on approach. Secured engine and pulled fire bottle. Fire put out on ground. Ruptured fuel line.
820705	Boeing	B-747-121	409	M	APAX	Pilot experienced fire warning on engine. Returned to airport. Fire damage to wiring insulation and cowling.
820707	Boeing	B-747-251B	343	N	APAX	On rotation at takeoff, aircraft had a fire warning on engine. Engine shut down. Fuel dumped. Fuel nozzle coking.
820710	Doug	DC-9	2	M	APAX	Pilot returned when engine fire warning came on. Fired both bottles. Damage to turbine section.
820821	Doug	DC-10-30F	393	M	APAX	During rotation, engine failed and fire warning came on. Passengers deplaned on taxiway. Landed overweight.
820830	Lkheed	L-1011	156	M	PXCG	Minor fire in lavatory. Extinguished fire immediately. Cause of fire unknown. FBI notified.
820907	Doug	DC-9-32	104	M	APAX	Crew extinguished fire in rest room. Damage limited to Kleenex box fixture. FAA security met aircraft on arrival.

820914	Boeing	B-747-212B	491	S	PXCG	Pilot experienced a sudden engine thrust loss. Had fire warning indications. Engine shutdown. Fire bottle used.
821106	Boeing	B-747-136	2	N	APAX	Oven A-3 caught on fire at cruise altitude. Overrun circuit breaker was opened and fire stopped.
821222	Boeing	B-747	242	M	PXCG	#4 engine caught on fire. Fire extinguished. Found #3 bearing breather manifold broken.
830301	Airbus	A-300-B2K3C	90	N	APAX	#1 engine temperature climbed to 957 degrees. Flight attendant reported fire. Fire bottle discharged. Returned.
830401	Boeing	B-727-223	3	N	APAX	Fire occurred in a bag under seat. Attendant put out fire. Fire resulted from a cigarette ash. Bad fire extinguisher.
830521	Boeing	B-747-123	3	N	PXCG	Fire warning light came on. Engine shut down. Leaking fuel ignited. Small fuel leak at the pylon connector.
830527	Lkheed	L-1011-3851	78	M	APAX	Heard a loud pop. Smoke and flames seen behind panel. Firebottle extinguished fire. Windshield heat breakers popped.
830602	Doug	DC-9-32	46	D	APAX	Fire started from unknown source in aft lavatory. Emergency landing was made and aircraft burned.

Date	Make	Model	TOB	Damage	Flt/Type	Remarks
830611	Boeing	B-727-222	149	M	PXCG	(No entry)
830705	Doug	DCB	137	S	PXCG	Experienced failure and fire of #1 engine on climbout. Engine shut down. Turbine failure and evidence of fire.
830712	CVAC	CVAC-440	8	M	PXCG	Engine fire on climbout. Engine shut down. Fire bottle used. Accessory section showed evidence of fire.
830721	Doug	DC-10-10	266	N	PXCG	Had a paper fire in waste chute in aft lavatory. Dumped water into chute. Found paper match and scorched Kleenex.
830808	Lkheed	L-1011	3	N	PXCG	Had a fire in lavatory. Attendant used fire extinguisher. Cigarette butt found among the paper seat covers.
830819	Boeing	B-747-123	401	N	PXCG	Saw smoke in lavatory. Believed a cigarette caused the fire. Water used to put fire out.
830822	Lkheed	L-1011	3	N	APAX	Used fire extinguisher to extinguish a burned food spill in galley oven.

830907	Boeing	B-727-23	41	N	PXCG	Reported electrical fire at cruise altitude. Situation under control. Found food left in one of the ovens.
830923	Boeing	B-727-200	92	N	PXCG	#1 engine fire on takeoff climb. Used fire bottles. Returned to DFW. Found a bleed air leak.
831017	BAC	BAC-146-200A	37	S	APAX	#3 engine caught on fire shortly after takeoff. Fire bottles used. Engine shutdown. Returned.
831207	Doug	DC-10	159	M	APAX	Blew a tire on landing rollout. Fire in right wheel well. Found right forward main gear truck had failed.
831214	Boeing	B-727-23	3	M	PXCG	Smoke and fire in a seat. Fire extinguished. Found empty cigarette package and napkin stuffed between seat cushion.
840201	Doug	DC-9-31	66	N	PXCG	Fire in the commode bowl. Taxiing for takeoff. Flames were from white tissue. Found smoke deposit under toilet seat.
840210	Boeing	B-747-136	126	M	APAX	#2 engine shut down due to inflight fire. Found #2 engine fuel shutoff valve in #2 fuel tank had failed.
840414	Boeing	B-747	3	M	ACAR	Engine fire warning light came on. Fire bottles used. Engine shut down. Fuel leak in right raceway.

Date	Make	Model	TOB	Damage	Flt/Type	Remarks
840419	Boeing	B-727-200	85	M	APAX	Explosion in engine and loss of hydraulic system on takeoff. Explosion came from engine combustion section.
840517	Boeing	B-727-2M7	142	N	APAX	Fire warning light came on. Fire bottles discharged. Aircraft evacuation via slides on taxiway.
840524	Boeing	B-727-100	7	N	APAX	Engine fire on climb to cruise. Found a stuck open starter valve. Allowed hot bleed air into the starter system.
840602	Boeing	B-747-121	182	N	PXCG	Low oil pressure on #3 engine. Fire warning. Fire bottles used. Evidence of fire in gear box. Cause undetermined.
840612	Boeing	B-727-200	3	S	PXCG	Flame and smoke coming from #3 engine. Continued to destination. Found starter failure. Cause undetermined.
840716	Doug	DC-10-10	227	M	PXCG	Inflight fire in forward lower galley. #4 oven. Firemen came on board at end of runway. Cause undetermined.
840806	Doug	DC-9	85	S	APAX	Decrease of EPR, loud bang, engine failed. Top cowling absent. Fuel nozzle lockring misassembled.

841115	Boeing	B-727	140	M	APAX	Flight attendant discovered small fire in aft lavatory. Extinguished.
850306	Boeing	B-727-223	124	M	APAX	Attempted restart on APU. Flames shot up outside aircraft. Some passenger panic. Improper restart procedure APU.
850503	Boeing	B-727-22	106	M	PXCG	Smoke coming from rear toilet. Extinguisher used and smoke dissipated. Found burned flush motor.
850511	Boeing	B-747-122	199	N	PXCG	Inflight fire warning #1 engine. Shut down and returned to airport. Evidence of small fire near #4 heat shield.
850620	Boeing	B-747-132	373	M	PXCG	Loss of oil and fire warning #2 engine. Shut down and diverted. Oil filter bearing failure. Fuel filter break, fire.

TABLE B-2

Cabin Smoke

Date	Make	Model	TOB	Damage	Flt/Type	Remarks
800220	Boeing	B-737	94	N	APAX	Flight taxied out. Had fuel smell in cockpit. Taxied back to gate. Departed 13 minutes later.
800222	Doug	DC-9-15	84	N	APAX	Flight returned due to odor and smoke. Cabin crew began evacuation procedures in error.
800228	Boeing	B-727-200	28	N	APAX	Malfunction in APU during start. Filled cabin with smoke. Cause unknown.
800311	Doug	DC-9	3	N	APAX	After takeoff, crew detected fuel fumes in cabin cockpit. Flight returned. Fuel spillage during refueling.
800413	Boeing	B-727	105	N	APAX	Smoke noted in cockpit. Flight returned to Austin. Interior lighting ballast transformer failed.
800417	Boeing	B-737-247	88	N	APAX	Pilot smelled smoke. Fuel indicator shorted and failed.
800514	Frchld	FH-227	36	N	APAX	Cabin blower failed, causing smoke in cockpit. Safe unscheduled landing made.

800615	Doug	DC-9-31	74	N	APAX	Flight diverted due to smoke in galley. Galley work light ballast had shorted.
800721	CVAC	STCAPJC	29	N	APAX	Flight aborted due to smoke in cockpit. Found faulty cabin compressor.
800918	Boeing	B-727	64	N	APAX	Taxied back to gate due to smoke in cockpit. Replaced radar unit.
801023	Doug	DC-9-51	81	N	APAX	Captain and first officer experienced nausea and disorientation from oven fumes. Made unscheduled stop.
801117	Boeing	B-727	108	N	APAX	Smoke in the cockpit of unknown origin, made an unscheduled landing. Turned out to be spoiler actuator motor.
801205	Boeing	B-727-227	57	M	APAX	Several breakers opened and crew smelled smoke after takeoff. Returned.
801223	Doug	DC-10-30F	111	N	APAX	Flight attendant notified crew of smoke in area of seat of row 15. Electrical system malfunction.
810113	Boeing	B-727	33	N	APAX	Sharp bump, gunpowder odor in cabin, precautionary landing. No bombs. Cause unknown.
810128	Doug	DC-10	83	M	APAX	Fire occurred in passenger carry on hand bag probably due to careless smoking.

Date	Make	Model	TOB	Damage	Flt/Type	Remarks
820109	Boeing	B-737-2H4	55	N	APAX	Enroute, half annunciator and master caution lights came on, replaced generator and APU module NR 693731423.
820119	Lkheed	L-1011	126	N	APAX	On takeoff, flight crew smelled smoke, donned masks. Smoke dissipated. Suspect deicing fluid in air conditioning.
820129	Doug	DC-10-10	180	N	APAX	Spray deicing fluid during deicing into APU inlet. Smoke fumes entered cabin. Passenger evacuated.
820208	Boeing	B-727-222	2	N	PXCG	Loud thump on takeoff. Landed normal at destination. No discrepancies.
820209	Doug	DC-9-31	42	N	APAX	Smoke in cabin from failed oil seal bearing, PN 20495054, on air cycle motor in tail.
820210	Doug	DC-9-32	2	N	PXCG	No right main gear down light. Landed OK. Pinned gear. Found defective "gear down lock" micro switch.
820309	Boeing	B-747	210	N	APAX	Smoke and electrical odor noticed in cockpit. Emergency equipment alerted. Dimming switch card had shorted.

820419	Doug	DC-9-80	15	S	APAX	Burning odor aft cabin. Returned to gate, off loaded. Found scorched paper towel on hot plate.
820430	Boeing	B-727-200	113	N	PXCG	Smoke in cabin on final approach. Evacuated passengers on landing. Radio problem with rescue unit. Air inlet door actuator.
820503	Boeing	B-727-223	47	N	PXCG	Unsafe gear indication. Pinned gear. Smoke in cabin from overheated pack. Inoperative gear accessory unit.
820517	Boeing	B-727-231	74	M	APAX	Smoke and fumes from galley in cabin. Shut down galley. After landing found a small piece of wood in oven.
820701	CVAC	STC-240T	35	N	APAX	Smoke appeared in cockpit after takeoff. Returned. Found EDC internal oil seal leaking into ventilating air.
820720	Boeing	B-727	5	N	APAX	Smoke in cockpit. Battery charged circuit popped. Battery charger overheated. PN 607013.
820810	Boeing	B-727-223	117	M	APAX	In climbout, rudder press-to-test light, PN054-1011-016, stuck, overheated, caught fire.
820908	Boeing	B-737-100	73	N	PXCG	Pilot detected strong smell of electrical overheat odor. Faulty transformer rectifier found. Unit replaced.

Date	Make	Model	TOB	Damage	Flt/Type	Remarks
820923	Boeing	B-727-223	117	N	APAX	Stewardess smelled electrical smoke in galley area. Found food spilled in oven. Cleaned oven. Checked ok.
820924	Doug	DC-10-10	239	N	APAX	Failure of non-reversible rudder hydraulic pump caused fumes to enter cabin at start up. Passengers deplaned.
820930	CVAC	STCAPJC	3	N	PXCG	Coffee maker shorted out creating smoke in cabin. Crew shut off the circuit breaker after landing.
821017	Doug	DC-9-81	127	M	APAX	Taxiing to gate ATC advised smoke coming from engine. Aircraft evacuated. Broken hydraulic line from reverse system.
821111	Doug	DC-8	196	N	PXCG	Crew detected smoke in cabin. Returned. Found a wing flood light transmitter in panel had overheated.
821226	Boeing	B-727-100	88	N	APAX	Aircraft filled with smoke. Crew on oxygen. Depressurized, smoke eliminated. Deicing fluid in water separator bag.
830110	Boeing	B-727	64	N	APAX	Fire warning #2 engine. Smoke in cabin. Passengers deplaned on taxiway. No evidence of fire. Broken stage duct.

830207	Boeing	B-757-200	108	N	APAX	During deicing of aircraft glycol fumes were emitted into cabin and cockpit. Fumes came through apu air inlet door.
830202	Boeing	B-727-22	110	N	PXCG	Cabin pressure uncontrollable. Smoke in cabin. Pressure regained at FL 180. Replaced pressure flow control valves.
830523	Boeing	B-737-222	59	N	APAX	Smoke in cabin while taxiing. Smoke came from an overheated air conditioning pack. Malfunctioning battery bus relay.
830527	Lkheed	L-1011-3851	78	M	APAX	Heard a loud pop. Smoke and flames seen behind panel. Firebottle extinguish fire. Windshield heat breakers popped.
830603	Boeing	B-727-200	131	N	APAX	Smoke in cabin. Aircraft evacuated at gate. Broken hydroline caused skydrol to vaporize. Discharge blew in cabin.
830617	Boeing	B-727-123	85	N	PXCG	Failure of the galley ovens overheat thermostat caused crew to suspect a fire. Smoke caused by food spillage.
030702	Frchld	FH-227C	38	M	APAX	Left DC generator failed at cruise altitude. Generator control panel and voltage regulator were malfunctioning.

Date	Make	Model	TOB	Damage	Flt/Type	Remarks
830716	Frchld	FH-227	28	N	APAX	Smelled smoke in cockpit. High amp draw on generator. Bearing failure in air condition air recirculating fan.
830721	Doug	DC-10-10	266	N	PXCG	Had a paper fire in waste chute in aft lavatory. Dumped water into chute. Found paper match and scorched Kleenex.
830819	Boeing	B-747-123	401	N	PXCG	Saw smoke in lavatory. Believed a cigarette caused the fire. Water used to put fire out.
830820	Boeing	B-737	2	N	APAX	Heard relays chattering. Saw light flashes behind breaker panel. Found transistors shorted out in card M562.
830923	Boeing	B-727	103	N	APAX	Cabin filled with smoke while holding for gate. Generator control bus circuit breaker was popped. Fans not working.
831007	Boeing	B-727-200	83	M	APAX	Smoke and electrical smell noticed on climbout. Replaced right hand taxiway turnoff light switch and burned wire.

831102	CVAC	CVAC-440	18	N	APAX	Cabin and cockpit became filled with smoke. Returned. Found cabin compressor bearing and seal damaged.
831117	Doug	DC-8-61	247	N	APAX	Smoke in cabin. Aborted takeoff roll. Two faulty fluorescent light ballasts caused the smoke. Tires smoking.
831124	Boeing	B-737-222	36	N	APAX	Smoke in cabin during start up. Landing gear squat switch open. Stopped ground cooling fans. Overheated right pack.
831201	Boeing	B-727-23	68	M	APAX	Smoke in cabin. Found overheated oven. Made precautionary landing en route. No damage found.
831214	Boeing	B-727-23	3	M	PXCG	Smoke and fire in a seat. Fire extinguished. Found empty cigarette package and napkin stuffed between seat cushion.
831226	Doug	DC-9	2	N	APAX	Suffered smoke in cabin at cruise altitude. Landed at Kansas City. Smoke cleared on final approach. Cause undetermined.
840105	Boeing	B-737-2B7	43	N	APAX	Smoke filled cabin. Slide evacuation. Crew not notified of deicing. Deicing fluid ingested into the air packs.

Date	Make	Model	TOB	Damage	Flt/Type	Remarks
840129	Boeing	B-727-23	122	N	APAX	Electrical power failed while taxiing. Haze in cabin. Engineer used placarded apu. Overheated air conditioner pack.
840212	EMB	EMB-110P1	11	N	APAX	Smoke in cabin after takeoff. Returned. Ground blower fan motor had shorted and burned up.
840215	Boeing	B-727-2M7	151	M	PXCG	APU field relay tripped. Lost all power. Pack overheated. Filled cockpit and cabin with smoke. Aircraft evacuated.
840419	Boeing	B-727	62	M	APAX	Smoke in cabin during enplanement of passengers. Found smoke was coming from a ballast for a fluorescent light.
840517	Doug	DC-10-40	248	M	PXCG	Smoke in cockpit. Returned. Found a wire bundle under copilot's glare shield burned through.
840627	Boeing	B-737-130	123	M	APAX	Cockpit had electrical smoke due to failure of capacitor in generator control unit. Antiskid inoperative, tires blew.
840731	Fokker	F-27	22	N	APAX	Smoke in cockpit from instrument panel. Turned off DME and cleared smoke. Weather radar had electronic short.

840806	Boeing	B-727-231	3	N	APAX	Smoke in cockpit. Crew went on oxygen. Found the weather radar unit had failed when its transformer had overheated.
840811	Boeing	B-727-100	56	N	PXCG	Smoke in cockpit. Masks not used. Found hi intensity light switch shorted.
840830	Fokker	F-27	26	N	PXCG	Smoke coming from aft galley water heater. Found a loose charred wire to water heater.
841112	Nihon	YS-11	4	N	PXCG	Smoke in cabin. Requested lower altitude. Released pressurization to air cabin. Emergency hydraulic pump had failed.
841201	Boeing	B-727-222	103	N	APAX	Smoke from cabin light panel. CB pulled. Light ballast defective.
841204	Boeing	B-727-2M7	89	M	APAX	Smoke and arcing from P611 panel. Aborted takeoff. Heat and smoke damage near left ground blower circuit breaker.

Date	Make	Model	TOB	Damage	Flt/Type	Remarks
850121	Boeing	B-727-243	2	N	APAX	Returned for landing due to smoke and fumes in cockpit. Skydrol in bleed air system, check valve failed.
850213	Doug	DC-10-40	140	N	PXCG	Pungent electrical odor, whiff of smoke seat 12. Multiplex encoder overheated.
850220	StBros	SD-360	3	N	OTHR	Smoke in cockpit after turning on ground power. External power relay damaged by electrical fault.
850225	Doug	DC-9-15	2	N	APAX	After takeoff aircraft filled with smoke. Uneventful unscheduled landing. Left and right coalescer bags replaced.
850309	Boeing	B-727-247	32	N	APAX	Smoke in passenger cabin. Overheated transformer at upper strobe light. Oil leaked out of transformer.
850408	Doug	DC-9-82	2	N	APAX	Circuit breakers for transformer rectifiers opened. Other breakers opened. Replaced rectifiers, AC control panel.
850503	Boeing	B-727-22	106	M	PXCG	Smoke coming from rear toilet. Extinguisher used and smoke dissipated. Found burned flush motor.

850522	Boeing	B-747-127	504	N	PXCG	Trash fire in container between lavatories. Used extinguishers and eliminated smoke.
850608	Boeing	B-727-247	68	N	APAX	Irritating fumes in cabin area. Passengers exited rear door. Flap control valve in left wheel well leaking into APU.
850626	Boeing	B-727-222	68	N	APAX	Cockpit and cabin filled with acrid smoke. Returned for landing, oil seal failed in air cycle machine.
850822	Doug	DC-9-82	131	N	APAX	Smoke and fumes in cockpit and cabin. Diverted where maintenance replaced oven in galley. Timer was defective.

TABLE B-3

Emergency Descent and Deployment of Oxygen Masks

Date	Make	Model	TOB	Damage	Flt/Type	Remarks
790211	Doug	DC-10-10	229	M	APAX	Faulty number 1 TR unit and bus control panel caused overhead warning lights to come on. D1 breaker popped.
790622	Doug	DC-9-31	29	N	APAX	Flight descended and returned after pressurization loss. Found broken line.
791021	Doug	DC-9-31	92	N	APAX	Crew descended from FL 310 and completed flight at 240. Found right air conditioning pack turbine failure.
791209	Doug	DC-9-31	60	N	APAX	Could not control cabin pressure at altitude. Maintenance could not duplicate the problem.
800305	Boeing	B-747	111	N	APAX	Flight had to shut down #3 engine due to compressor stalls. Metal residue found in bleed air ports.
800307	Doug	DC-9	91	N	APAX	Cabin pressure started to climb. Maintenance unable to duplicate problem.
800323	Frchld	FH-227	38	N	APAX	Pilot unable to get "gear safe" light. Only gear in transit. Made normal landing, problem was in switch.

800328	Boeing	B-727	78	N	APAX	Engine shut down due to turbine failure. Flight returned and cancelled.
800329	Boeing	B-707	108	N	APAX	Lost hydraulic fluid in flight. Worn 0 ring on reservoir drain plug.
800406	Boeing	B-740-U	111	M	APAX	Aircraft sustained bird strike causing damage to radome and forward bulkhead.
800406	Boeing	B-747	0	N	APAX	Aft cargo fire warning came on flight. Maintenance replace faulty fire protection system.
800515	Boeing	B-727-22	40	N	APAX	Cabin pressurization could not be controlled, oxygen masks deployed. Safe landing made after emergency descent.
800515	Doug	DC-9-14	2	N	APAX	Flight diverted after cabin pressurization climbed. Replaced radio rack venturi shut off valve.
800531	Boeing	B-747	111	N	APAX	Engines developed power problems at altitude but application of fuel engine heat cleared the problems.
800629	Boeing	B-727	105	N	APAX	Pressurization problems at FL 370. Replaced door seal outflow valves and pressurization controllers.
800703	Doug	DC-9	63	N	APAX	Flight had uncontrollable rise in cabin pressure; craft descended to 14,000 and masks deployed.

Date	Make	Model	TOB	Damage	Flt/Type	Remarks
800711	Boeing	B-727	97	N	APAX	Emergency descent when control of cabin pressure lost at FL 260. Door seals outflow valve leaking.
800711	Boeing	B-727	97	N	APAX	Emergency descent when control of cabin pressure lost at FL 260. Door seals outflow valve leaking.
800715	Boeing	B-727-247	106	N	APAX	Experienced rapid decompression due to faulty pressure controller.
800731	Boeing	B-727-200	87	N	APAX	Lost cabin pressure on climbout. Returned to airport.
800818	Boeing	B-727-200	88	N	APAX	Hydraulic "A" system failure due to failed bolts on flap selector valve.
801104	Doug	DC-9	18	N	APAX	Aircraft experienced depressurization at FL 330. Masks dropped. Emergency descent. Landed uneventfully.
801126	Boeing	B-727-30	79	N	APAX	Pressurization lost at 17,800 feet. Rear cargo door not properly latched.
801222	Boeing	B-747-121	16	N	APAX	Emergency descent after crew unable to control pressurization. Replaced auto controller.

801227	Doug	DC-9-31	115	N	APAX	Flight experienced electrical failure and depressurization. A wire had broken in the transformer.
810106	Boeing	B-727	76	N	APAX	Cabin altitude started to climb. Descended to safe altitude. Found cracked starter duct.
810110	Doug	DC-9-32	64	N	APAX	Cabin lost pressurization. Caused descent emergency. Replaced right generator, control panel and auto relay system.
810310	Boeing	B-727-227	68	N	APAX	Cabin started to climb at Flight Level 270. Crew could not control. Cabin pressure controller replaced.
810430	Doug	DC-9	52	N	APAX	Aircraft made an emergency descent due to loss of air flow and pressure rising.
810515	Doug	DC-9	72	N	APAX	Cabin began to climb while cruising at 35,000 ft. Declared emergency and descended to 10,000 ft. Duct found blown off.
810617	Lkheed	L-1011-3851	2	N	APAX	Lost pressurization at FL 350. Cargo door seal failed.
810717	Boeing	B-727-223	69	N	APAX	Losing pressurization emergency descent and landing. System to be checked.

Date	Make	Model	TOB	Damage	Flt/Type	Remarks
810719	Lkheed	L-1011	236	N	APAX	Windshield cracked en route. Descended to unpressurized altitude. Landed destination. Replaced windshield.
810817	Doug	DC-8-63	4	N	ACAR	#2 engine fire warning indication. Made emergency landing. Found broken wire in system. Repaired wire.
810904	Boeing	B-707-323B	86	N	APAX	Minor problem crack in outflow valve. Flight engineer then mismanaged cabin controls. Caused cabin decompression.
811108	Doug	DC-9-31	95	N	APAX	Pressurization problem, emergency descent, some oxygen masks failed to deploy, ferried to maintenance.
811124	Boeing	B-727-227	66	N	PXCG	Cabin pressure climbed, emergency descent, deployed masks. Engine bleed valve and diode in aural horn replaced.
811201	Doug	DC-9	0	N	APAX	Cabin altitude problems, descent emergency, oxygen mask deployed. Cabin outflow valve jammed open by dirt.
811217	Doug	DC-8-61	230	M	APAX	Rapid cabin pressure loss, emergency descent, returned. Service air condition panel separated. Check valve jammed.

82000827	CVAC	STCAPJC	21	N	APAX	Aircraft lost cabin pressure at cruise altitude. Found broken wire at left hand squat switch.
820111	Doug	DC-8-63	229	N	APAX	Uncontrolled cabin pressure. Passengers, attendants not advised. Burst water tank frozen outflow valve.
820327	Boeing	B-727-231	143	N	APAX	Cabin altitude started climbing. Pressure stabilized by closing engine bleed valve.
820517	Doug	DC-9	83	N	PXCG	Cabin lost pressure, buffeting on emergency descent with speed brake and gear out. Cause of problems unknown.
820627	Doug	DC-8-61	162	N	PXCG	Heard rumble and simultaneously cabin depressurized. Outlet duct parted at a connection. Had a compressor vibration.
820728	Boeing	B-727-223	151	N	PXCG	Takeoff warning horn activated due to misaligned APU. Cargo outflow valve stuck. Depressurization unnoticed by crew.
820913	Boeing	B-727	37	N	APAX	#3 bleed air tripped at 37,000 feet. Used emergency procedures. Cabin pressure climbed, masks dropped.
821221	Doug	DC-10-10	122	N	PXCG	Unable to maintain cabin pressure. Emergency descent to 1000 feet. Found a defective cabin pressure controller.

Date	Make	Model	TOB	Damage	Flt/Type	Remarks
830107	Boeing	B-727-225	3	M	PXCG	Unable to control cabin pressure. Emergency descent to 10,000 ft. Found rupture in right pack ducting behind APU.
830121	Boeing	B-727-223	119	N	APAX	A blown duct and water separator in pressurization system caused rapid cabin altitude climb. Made emergency descent.
830131	Doug	DC-9-14	75	N	PXCG	Lost control of pressurization system. Regained control at 1400 feet. Replaced cabin pressure controller.
830212	Boeing	B-737-3H4	112	N	APAX	Outside window panel blew out. Descended 10,000 ft. Replaced shattered window and two cracked windows.
830505	Boeing	B-747-151	398	M	APAX	Fiberglass honeycomb panel between fuselage and pylon on left wing cracked. Had light turbulence. Fuel dumped.
830601	Boeing	B-727-730	3	N	APAX	Lost complete pressurization at cruise altitude. All masks deployed. Found bird nest jammed in valve butterfly.
830612	Boeing	B-727-295	83	N	APAX	Smoke in cabin. Made emergency descent. Found left air conditioning pack overheated.

830821	Doug	DC-10-10	254	N	APAX	Lost all pressurization at cruise altitude. Unable to control cabin pressure. Landed Minneapolis.
830914	Boeing	B-727-224	97	N	APAX	Cabin pressure failed at FL 30. Emergency descent made. Malfunction in cabin pressure control unit.
830916	Doug	DC-9-31	53	N	APAX	Auto pressure failed causing a rapid decompression. Found a malfunctioning controller.
831025	Boeing	B-727-200	138	N	APAX	Right pack tripped. Executed emergency descent to lower altitude. Replaced A.C.M. 250 degree overheat switch.
831110	Swrngn	SA-227	12	N	APAX	Experienced decompression on descent. Crew and passenger oxygen turned on. No malfunctions found.
831116	Boeing	B-727-222	92	N	APAX	Lost pressurization. Regained pressurization at 14,000 ft. Four masks failed to operate. Code 7700 used on descent.
831124	Doug	DC-9	3	N	APAX	Passenger became ill. Emergency descent into El Paso. Met by paramedics and taken to a medical facility.
831222	Boeing	B-727-22	2	N	PXCG	Automatic control of pressurization became inoperative. Replaced manual pressure controller, pack valves, and switch.

Date	Make	Model	TOB	Damage	Flt/Type	Remarks
840123	BAC	BAC-111-204	63	N	APAX	Experienced rapid depressurization at altitude. Air conditioning ground access cover assembly had worked loose.
840210	Boeing	B-727-222	88	N	APAX	Had cabin depressurization at cruise altitude. Replaced auto and manual cabin pressure controller.
840218	Doug	DC-9	37	N	APAX	Cabin pressure problem. Emergency descent. Replaced amplifier, actuator and controller of pressurization system.
840229	Doug	DC-8-61	245	N	PXCG	Emergency descent due to loss of pressurization. Found external air conditioning door was torn off.
840307	Boeing	B-737-100	35	N	APAX	Lost cabin pressure. Emergency descent. Door seal of aft air stair door had become dislodged from its track.
840310	Boeing	B-727-23	41	M	PXCG	Rapid decompression at cruise altitude. Found fatigue failure of the fuselage skin forward of the aft cargo door.
840408	Doug	DC-9-31	101	N	PXCG	Unable to control cabin pressure. Descended to a lower altitude. Replaced the pressure amplifier.

840427	Doug	DC-9-80	39	N	APAX	Lost pressurization control at cruise altitude. Cause undetermined.
840527	Boeing	B-727-222	128	N	PXCG	Pressurization system failed during descent. Replaced the automatic pressure controller and left outflow valve.
840623	Boeing	B-727-223	2	N	APAX	Engine flamed out at cruise altitude. Found a faulty fuel flow transmitter leaking fuel. Replaced fuel transmitter.
840715	Lkheed	L-1011-3851	181	N	APAX	Bomb threat received. Landed Orlando. Went to remote area for evacuation and search. Evaluated via slide. No bomb.
840727	Doug	DC-9-80	2	N	APAX	Loss of pressurization required rapid descent. Outflow valve and number 1 cabin pressure controller replaced.
840808	Doug	DC-9	2	N	APAX	Pressurization warning light came on. Dropped oxygen masks. No loss of pressure. Found cargo door latch faulty.
840808	Boeing	B-727	3	N	APAX	Encountered brief large hail in clouds at FL 330. Cracked windshield. Made descent to VFR [visual flight rule conditions] at FL 270. No injuries.

Date	Make	Model	TOB	Damage	Flt/Type	Remarks
840813	Doug	DC-10-10	249	N	APAX	Unable to control pressurization. Found NR 2 pack duct loose and sense line broken.
840915	Doug	DC-10-10	3	N	APAX	Flight officer became unconscious at his duty station. Recovered by use of oxygen. Cause undetermined.
840502	Doug	DC-9	61	N	PXCG	Loss of pressurization at cruise altitude. Changed the outflow valve actuator and differential pressure sensor.
850107	Frchld	FH-227	31	N	APAX	While en route left oil pressure light illuminated. Engine shutdown. Oil blowing out of breather. Replaced engine.
850206	Doug	DC-9-32	57	N	PXCG	Slow decompression en route. Descended until manual operation was obtained. Changed pressurization controller.
850306	Doug	DC-9-32	40	N	APAX	Cabin rate of climb went to max rate. Emergency descent. Control of pressure regained. Flexible coupling failed.

850512	Boeing	B-727-223	151	M	APAX	Pressurization problems during climb. Emergency descent. Replaced water separator, temp controller, sensor, valve.
850608	Boeing	B-737	115	N	APAX	Pressurization system, auto fail system did not work. Diverted. Cause of malfunction not reported.
850612	Doug	DC-9-32	3	N	PXCG	Depressurized at FL 330. Emergency descent. Passenger fainted. Possible oxygen system malfunction.
850623	Doug	DC-9-80	123	N	PXCG	Decompression occurred at FL 350. Diverted and descended. Maintenance replaced selector, controller, valve actuator.
850923	Boeing	B-727-222	109	N	PXCG	Uncontrollable increase in cabin pressurization. Emergency descent and safe landing. Recurrent undetermined problem.

Appendix C

Airliner Cabin Safety Regulations and Standards

This appendix presents information about airliner cabin safety regulations, standards, and recommendations. It is based on items listed in the Cabin Safety Index prepared by the FAA Civil Aeromedical Institute,[12] supplemented by relevant items in the Federal Register since the index was published. It presents relevant regulations and recommendations concerning emergency procedures, nonemergency procedures, equipment, crew training, and passenger information and briefing with respect to fires (Table C-1), decompression (Table C-2), medical emergencies (Table C-3), and ditching and evacuation (Table C-4). Table C-5 deals with preflight and in-flight announcements, and Table C-6 presents a summary of typical air carrier operating procedures with respect to firefighting and firefighting training.

TABLE C-1 Standards, Regulations, and Recommendations about Fires

Emergency Procedures

Air Carrier Operations Bulletin: Air carrier emergency procedures pertaining to lower-lobe operation should be reviewed.[22]

Nonemergency Procedures

Regulation: No passenger or crew member may smoke while the "no smoking" sign is lighted, and each passenger shall fasten his or her seat belt and keep it fastened while the seat belt sign is on.[11]

Airworthiness Directive: 1,000-h periodic inspections, and repairs as necessary, of all lavatory trash receptacles to ensure fire containment procedures.[21]

Air Carrier Operations Bulletin: Inspection of lavatory before takeoff and periodically during flight.[20]

Equipment

Regulation: Hand fire extinguishers available for all baggage compartments with access by crew members.[8]

Regulation: Hand fire extinguishers available for crew, passenger, and cargo compartments,[6] uniformly distributed in passenger compartments with two Halon 1211 extinguishers per airplane.[15]

Regulation: Protective breathing equipment must be installed for each isolated separate compartment in the airplane, including upper-and lower-lobe galleys.[13]

Proposed regulation: Protective breathing equipment that protects crew members from effects of smoke, carbon dioxide, or other harmful gases and that protects crew members while combatting fires on board; one such device must be in each upper-or lower-lobe galley, one on the flight deck, one for use in each accessible cargo compartment, and one within 3 ft of each required fire extinguisher.[25]
Regulation: Floor proximity emergency escape-path marking for passengers when all sources of illumination more than 4 ft above the cabin aisle are obscured.[19]
Regulation: Smoke detectors in each lavatory and galley; automatic fire extinguisher for each lavatory trash receptacle.[15]
Airworthiness Directive: Installation of "no smoking" signs on each side of lavatory doors and ashtrays near lavatory entrances.[21]

<u>Crew Training</u>

Regulation: Instruction in emergency assignments and procedures; location, function, and operation of emergency equipment (i.e., portable fire extinguishers, including the type for different classes of fires); handling of fires on ground and in flight.[3]
Regulation: Actual operation of emergency equipment for each type of aircraft once each 24 calendar mo.[3]
Proposed Regulation: One-time emergency drill to be accomplished during initial training; additional emergency training to be accomplished once each 24 mo.[25]
Air Carrier Operations Bulletin: Initiate ground training or operations bulletins to inform flight deck crews and cabin crews of the causes, characteristics, and hazards associated with fluorescent light ballast fires.[17]

Air Carrier Operations Bulletin: Review emergency procedures pertaining to the lower lobe to ensure that procedures and equipment are adequate.[22]
Passenger Information/Briefing
Regulation: Preflight briefing concerning smoking.[2]
Airworthiness Directive: Preflight briefing not to smoke in lavatories.[21]

TABLE C-2 Standards, Regulations, and Recommendations about Decompression

Emergency Procedures (None)
Nonemergency Procedures
Regulation: Minimal mass flow of supplemental oxygen is specified in terms of mean tracheal oxygen partial pressure (precise specifications depend on exact equipment, altitude, duration at altitude, and other factors).[9]
Equipment
Regulation: Supplemental oxygen must be available for crew and passengers whenever the airplane is operated above 10,000 ft (exact provisions depend on the flight altitude and duration at altitude).[14]
Regulation: Each flight attendant shall, during flight above flight level 250 (25,000 ft), carry portable oxygen equipment with at least a 15-min supply of oxygen, unless enough units or spare outlets and masks are distributed throughout the cabin to ensure immediate availability to each cabin attendant.[14]
Crew Training
Regulation: Instruction in emergency assignments and procedures; location, function, and operation of emergency equipment; instruction in handling emergency situations (including rapid depressurization).[3]
Regulation: Crew members who serve in operations above 25,000 ft must receive instruction in respiration, hypoxia, duration of consciousness without supplemental oxygen at altitude, gas expansion, gas bubble formation, physical phenomena, and incidents of depressurization.[3]

Passenger Information/Briefing

Regulation: Before flight is conducted above flight level 250 (25,000 ft), a crew member shall instruct the passengers on the necessity of using oxygen in the event of cabin depressurization.[14]

TABLE C-3 Standards, Regulations, and Recommendations about Medical Emergencies

Emergency Procedures (None)

Nonemergency Procedures

Regulation: Conditions under which a passenger may carry and operate equipment for the storage, generation, or dispensing of oxygen are specified.[10]

Equipment

Regulation: Approved first-aid kits for treatment of injuries likely to occur in flight or in minor accidents must be provided (the number of kits varies according to the number of passengers carried).[6]

Regulation: Emergency medical equipment; one medical kit would be required on each passenger-carrying flight and should contain equipment and drugs required to provide basic life support during medical emergencies that might occur during flight, such as myocardial infarction, severe allergic reactions, acute asthma, insulin shock, protracted seizures, and childbirth.[18]

Crew Training

Regulation: Instruction in emergency assignments and procedures; location, function, and operation of emergency equipment (including first-aid equipment and its proper use); instruction in handling emergency situations (including illness, injury, or other abnormal situation involving passengers or crew members).[3]

Regulation: Familiarization with the emergency medical kit.[18]

Passenger Information/Briefing
Regulation: Crew members who serve in operations above 25,000 ft must receive instruction in respiration, hypoxia, duration of consciousness withouth supplemental oxygen at altitude, gas expansion, gas bubble formation, physical phenomena, and incidents of depressurization.[3]
Passenger Information/Briefing (None)

TABLE C-4 Standards, Regulations, and Recommendations about Ditching and Evacuation

Emergency Procedures
Air Carrier Operations Bulletin: In case of an unplanned emergency landing, the flight attendants might have only enough time to give a short command, such as "lean over" or "grab your ankles".[16]
Air Carrier Operations Bulletin: In case of a planned emergency landing, passengers should be briefed on proper bracing positions.[16]
Nonemergency Procedures
Air Carrier Operations Bulletin: Principal operations inspectors should ensure that flight attendants are fully aware that escape slides should be inflated manually if autoinflation fails.[23]
Air Carrier Operations Bulletin: Principal operations inspectors should evaluate seat spacing and passenger briefing card brace positions.[16]
Equipment
Regulation: Each passenger-carrying landplane emergency exit (other than over-the-wing) that is more than 6 ft from the ground with the airplane on the ground and the landing gear extended must have an approved means to assist occupants in descending to the ground.[1]
Regulation: An approved flotation means or a life preserver must be within easy reach of each seated occupant for extended over-water operation;[7] enough liferafts to accommodate all occupants must be provided;[5] each certificate holder shall demonstrate the effectiveness of emergency evacuation equipment and procedures and shall describe these in its manual.[4]

Crew Training

Regulation: Instruction in emergency assignments and procedures; location, function, and operation of emergency equipment, including, for ditching: cockpit preparation; crew coordination; passenger briefing and cabin preparation; donning and inflation of life preservers; removal and inflation of each type of liferaft; transfer of each type of slide/raft from one door to another; deployment, inflation, and detachment of each type of slide/raft; use of liferaft; boarding of passengers and crew into a raft or a slide/raft pack.[3]

Air Carrier Operations Bulletin: Principal operations inspectors must continually review their assigned air carriers' emergency evacuation procedures.[24]

Passenger Information/Briefing

Regulation: In extended over-water operations, all passengers are to be orally briefed on the location and operation of life preservers, liferafts, and other flotation means, including a demonstration of the method of donning and inflating a life preserver.[2]

TABLE C-5 Additional Passenger Briefing

Preflight Announcements
Regulation: Smoking; location of emergency exits; use of safety belts, including how to fasten and unfasten them; location and use of required emergency flotation devices.[2]
In-flight Announcements
Regulation: Immediately before or immediately after the seatbelt sign is turned off, an announcement shall be made that passengers should keep their seatbelts fastened while seated.[2]

TABLE C-6 Summary of Typical Air Carrier Operating Procedures with Respect to Firefighting and Firefighting Training

Firefighting
Cabin crew member discovering fire to identify the source and type of fire and switch off any electric supply involved, take nearest appropriate fire extinguisher, and attack the fire.
Second cabin crew member to be called to alert the captain and the senior member of the cabin crew.
Senior member of the cabin crew takes charge of firefighting, ensures that all necessary resources are available, and ensures that all portable oxygen bottles are removed from the scene.
Passengers are advised to keep heads down and to cover noses and mouths; if time and conditions permit, damp face cloths would be distributed.
Firefighting Training
On initial course, cabin crew are trained in the use of fire extinguishers and smoke protection hoods; practical use of equipment at the fire training ground is included; each student experiences a short period in a smoke chamber.
Students are required to demonstrate proficiency in firefighting in a synthetic smoke-filled cabin mockup.
Every third year, cabin fire and smoke is the main theme for flight crew and cabin crew annual checks; audiovisual review and familiarization with equipment under guidance of instructor; fire-smoke situation presented to cabin crew without warning in cabin mockup to check proficiency; flight crew briefed on cabin fire drill.

REFERENCES

1. Additional emergency equipment. Code of Federal Regulations, Title 14, Pt. 121.310. Washington, D.C.: U.S. Government Printing Office, 1985.
2. Briefing passengers before takeoff. Code of Federal Regulations, Title 14, Pt. 121.571. Washington, D.C.: U.S. Government Printing Office, 1985.
3. Crew member emergency training. Code of Federal Regualations, Title 14, Pt. 121.417. Washington, D.C.: U.S. Government Printing Office, 1985.
4. Emergency and emergency evacuation duties. Code of Federal Regulations, Title 14, Pt. 121.397. Washington, D.C.: U.S. Government Printing Office, 1985.
5. Emergency equipment for extended over-water operations. Code of Federal Regulations, Title 14, Pt. 121.339. Washington, D.C.: U.S. Government Printing Office, 1985.
6. Emergency equipment. Code of Federal Regulations, Title 14, Pt. 121.309. Washington, D.C.: U.S. Government Printing Office, 1985.
7. Emergency flotation means. Code of Federal Regulations, Title 14, Pt. 121.340. Washington, D.C.: U.S. Government Printing Office, 1985.
8. Fire precautions. Code of Federal Regulations, Title 14, Pt. 121.221. Washington, D.C.: U.S. Government Printing Office, 1985.
9. Minimum mass flow of supplemental oxygen. Code of Federal Regulations, Title 14, Pt. 25.1443. Washington, D.C.: U.S. Government Printing Office, 1985.
10. Oxygen for medical use by passengers. Code of Federal Regulations, Title 14, Pt. 121.574. Washington, D.C.: U.S. Government Printing Office, 1985.

11. Passenger information. Code of Federal Regulations, Title 14, Pt. 121.317. Washington, D.C.: U.S. Government Printing Office, 1985.
12. Pollard, D. W., J. A. Steen, W. J. Biron, and R. L. Cremer. Cabin Safety Subject Index. FAA-AM-84-1. Oklahoma City, Okla.: U.S. Federal Aviation Administration, Civil Aeromedical Institute, 1984.
13. Protective breathing equipment. Code of Federal Regulations, Title 14, Pt. 25.1439. Washington, D.C.: U.S. Government Printing Office, 1985.
14. Supplemental oxygen for emergency descent and for first aid; turbine engine powered airplanes with pressurized cabins. Code of Federal Regulations, Title 14, Pt. 121.333. Washington, D.C.: U.S. Government Printing Office, 1985.
15. U.S. Federal Aviation Administration. Airplane cabin fire protection: Final rule. Federal Register 50(29 March):12726–12733, 1985.
16. U.S. Federal Aviation Administration. Brace for impact positions. ACOB No. 1-76-23, para. 224. In Air Carrier Operations Bulletins, Consolidated Reprint. Washington, D.C.: U.S. Federal Aviation Administration, 1985.
17. U.S. Federal Aviation Administration. Cabin fluorescent light ballast fires. ACOB No. 8-82-1, para. 966. In Air Carrier Operations Bulletins, Consolidated Reprint. Washington, D.C.: U.S. Federal Aviation Administration, 1985.
18. U.S. Federal Aviation Administration. Emergency medical equipment: Final rule. Federal Register 51(9 Jan.):1218–1223, 1986.
19. U.S. Federal Aviation Administration. Floor proximity emergency escape path marking: Final rule. Federal Register 49(26 Oct.):43182–43186, 1984.

20. U.S. Federal Aviation Administration. In-flight lavatory fires. ACOB No. 1-76-17, para. 218. In Air Carrier Operations Bulletins, Consolidated Reprint. Washington, D.C.: U.S. Federal Aviation Administration, 1985.
21. U.S. Federal Aviation Administration. Inspections of Lavatory Trash Receptacles for Fire Containment Procedures. Airworthiness Directive 74-09-08, Docket No. 13603.
22. U.S. Federal Aviation Administration. Problems associated with lower lobe galleys. ACOB No. 1-76-14, para. 215(4). In Air Carrier Operations Bulletins, Consolidated Reprint. Washington, D.C.: U.S. Federal Aviation Administration, 1985.
23. U.S. Federal Aviation Administration. Procedures and training for emergency evacuations. ACOB No. 1-76-21, para. 222. In Air Carrier Operations Bulletins, Consolidated Reprint. Washington, D.C.: U.S. Federal Aviation Administration, 1985.
24. U.S. Federal Aviation Administration. Procedures and training for emergency evacuations. ACOB No. 8-76-10, para. 911. In Air Carrier Operations Bulletins, Consolidated Reprint. Washington, D.C.: U.S. Federal Aviation Administration, 1985.
25. U.S. Federal Aviation Administration. Protective breathing equipment: Notice of proposed rulemaking. Federal Register 50(Oct. 10):41452–41459, 1985.

Glossary

ACOB	Air Carrier Operations Bulletin.
ACGIH	American Conference of Governmental Industrial Hygienists.
ACM	See Air-cycle machine.
ACRE	Aircraft Radiation Exposure, a model of cosmic radiation exposure of aircraft passengers.
Aerosol	A suspension of liquid droplets or solid particles in a gas.
Air carrier	A person or group of persons using aircraft to transport persons, property, and mail.
Aircraft	A vehicle designed or used for flight.
Air-cycle machine	A turbine-compressor combination used to reduce air temperature by extracting energy from an air stream; part of the environmental control unit. (Abbr., ACM.)
Air-ex-change	Replacement of equivalent air volume in a compartment with fresh air.
Air-ex-change rate	Number of air-exchanges per unit time.
Air pack	See Environmental control unit.

Airplane A heavier-than-air, power-driven, fixed-wing aircraft that is supported by the dynamic reaction of air against its wings.

Airworthy Suitable for safe flight.

Angina pectoris Severe restricting pain in the chest, usually caused by insufficient blood flow to the heart muscle.

APU See Auxiliary power unit.

ASHRAE American Society of Heating, Refrigerating, and Air-Conditioning Engineers, Inc.

Auxiliary power unit A power unit that can be used in addition to the main sources of power. (Abbr., APU.)

Avionics Aviation electric and electronic equipment in the cockpit.

B-747-SP Special-performance model of the B-747 that is equipped with a catalytic converter to decompose atmospheric ozone; used for routes through ozone-laden portions of the atmosphere.

Background radiation Natural radiation in the environment, including cosmic radiation and radiation from naturally radioactive elements.

Bleed air Air from the compressor used for cabin ventilation.

Bypass ratio Flow ratio of low-pressure air in the fan to high-pressure air in the engine core.

CAB Civil Aeronautics Board (now defunct).

Cabin The sector of an aircraft occupied by passengers.

Cabin crew Flight attendants.

Carboxy-hemoglobin (COHb) Combination of carbon monoxide and hemoglobin; at high concentrations, carboxyhemoglobin interferes with the transfer of carbon dioxide and oxygen in the blood, causing asphyxiation.

Certificated route air carrier An air carrier holding a Certificate of Public Convenience and Necessity from the Department of Transportation, authorized to provide scheduled service over specified routes.

Certification The process by which FAA approves all air carriers, pilots, aircraft models, etc., to ensure compliance with applicable statutes and regulations.

cfm Cubic feet per minute.

CFR Code of Federal Regulations.

CO Carbon monoxide.

CO_2 Carbon dioxide.

COHb See Carboxyhemoglobin.

Commuter airline An air carrier that makes at least five scheduled round trips per week with small aircraft.

COPD Chronic obstructive pulmonary disease.

Cosmic radiation Energetic particles of extraterrestrial origin that strike the earth's atmosphere, as well as secondary particles generated by these interactions.

Depressurization Loss of cabin pressure during flight.

Disinsection	Use of insecticides to exterminate insect pests.
ECAC	European Civil Aviation Conference.
ECU	See Environmental control unit.
ECS	See Environmental control system.
Enplane-ments	The number of times that revenue passengers board flights; a passenger who changes from Flight A to Flight B en route to a destination counts as two enplanements.
Environ-mental control sys-tem	The total air-conditioning, heating, ventilation, and pressurization system on an aircraft, which provides occupants with a suitably controlled atmosphere to maintain comfort and safety; consists of several environmental control units. (Abbr., ECS.)
Environ-mental control unit	Equipment used to condition high-temperature, high-pressure air from a jet engine before delivery to the cabin; usually consists of an air-cycle machine and one or more heat exchangers. Also called air pack. (Abbr., ECU.)
Environ-mental tobacco smoke	Total air pollution due to burning of tobacco products, including sidestream and exhaled smoke. (Abbr., ETS)
ETS	See Environmental tobacco smoke.
FAA	Federal Aviation Administration.
FAR	Federal Air Regulation.
FEF	See Forced expiratory flow.
FEV	See Forced expiratory volume.
FEV_1	Maximal volume of air that can be exhaled in 1 s.

Flashover The point during a fire at which the temperature in a compartment becomes high enough for all materials and gases to ignite spontaneously.

Flight crew The pilots, navigators, engineers, and others needed to operate the aircraft.

Flight deck Cockpit area of an aircraft.

Flight level A level of constant atmospheric pressure related to a reference point of 29.92 in. of mercury; stated in digits that represent hundreds of feet, i.e., flight level 255 indicates a barometric altitude of 25,500 ft.

Floor proximity escape-route markers Illuminated exit signs near the floor designed to be visible in a smoke emergency.

Forced expiratory flow The average flow rate during forced expiration in a designated interval of the expiration period. (Abbr., FEF.) The interval is indicated by a subscript; e.g., $FEV_{25-75\%}$ refers to the average flow rate during the middle half of the expiration period.

Forced expiratory volume Maximal volume that can be exhaled in a specific period. (Abbr., FEV.) The period, in seconds, is indicated by a subscript, e.g., FEV_1.

Galley Food preparation area of an aircraft.

GAO U.S. General Accounting Office.

Gasper Individual air outlet usually placed in the ceiling above each seat, allowing the passenger to regulate the volume and direction of air flowing from the gasper to the seat.

Ground fumes Airport pollution, including emission from aircraft on the ground, maintenance vehicles, and airport transportation vehicles.

HVAC Heating, ventilation, and air-conditioning.

Hypoxia A condition resulting from a decrease in oxygen tension in the inspired air or a reduction in the oxygen-carrying capacity of the blood.

Load factor See Passenger load factor.

Lower lobe The part of an aircraft below the main floor of the cabin.

Mainstream smoke Smoke that a smoker inhales directly from a cigarette, or other tobacco product.

Makeup air Outside (fresh) air that is used in aircraft ventilation, which must be conditioned by heating, cooling, filtering, etc., before being delivered to occupied spaces.

Microbial aerosols A suspension of microorganisms in air.

mrem Millirem, 0.001 rem.

Myocardial infarction Sudden heart failure caused by interruption of blood supply to the heart muscle due to blockage of blood vessels or necrosis (death) of tissue in part of the heart due to this blockage.

Narrow-body aircraft An airplane with only one passenger aisle and generally fewer than 200 seats, e.g., B-727, B-737, B-757, DC-9-80, and BAE-146.

NO_2 Nitrogen dioxide.

Nonscheduled carriers Air carriers that provide charter services.

NTSB National Transportation Safety Board.

Offgassing Emission of low-vapor-pressure volatile organic vapors into the air, e.g., release of formaldehyde from urea-formaldehyde resin used to glue plywood.

OSHA Occupational Health and Safety Administration.

Outside air Air from outside the aircraft; outside air is mixed with air inside the aircraft, thereby diluting or "flushing" stale air to the outside.

Pack See Environmental control unit.

Part 121 airlines Certificated route air carriers that operate under the rules of Title 14, Part 121, of the Code of Federal Regulations.

Part 135 airlines Air carriers, primarily commuter airlines and air taxis, that operate under the rules of Title 14, Part 135, of the Code of Federal Regulations.

Partial pressure Pressure exerted by a single gas in a mixture of gases; commonly expressed in millimeters of mercury.

Passenger flight hours The number of passengers multiplied by the flight duration in hours.

Passenger load factor Percentage of aircraft seating capacity that is sold and used.

Pathogen Microorganism capable of causing disease.

pCO_2 Partial pressure of carbon dioxide.

PEFR Peak expiratory flow rate.

Plenum A common chamber in which air from different sources is mixed before being distributed to the cabin; the air can come from heating units, from the outside (fresh air), and from inside the aircraft (recirculated air).

Pneumothorax Presence of gas in the chest cavity outside the lungs.

pO_2 Partial pressure of oxygen.

ppm Parts per million.

Pressurization system The part of an aircraft's environmental control system that keeps cabin pressure relatively constant, not exceeding the legal maximal equivalent altitude of 8,000 ft.

Protective breathing device A device worn over the nose and/or mouth that allows the wearer to breathe relatively clean air for a short time in the presence of smoke and toxic fumes.

Rad The unit of absorbed dose of radiation equal to 100 ergs/g.

Recirculation air Air that is reused for aircraft ventilation after being removed from the cabin; it is usually filtered to remove particles, aerosols, and gaseous tars from tobacco smoke and is usually diluted with fresh air before being returned to the cabin.

Relative humidity The amount of moisture in air compared with the maximal amount that the air could contain at the same temperature; expressed as a percentage. (Abbr., RH.)

rem Roentgen equivalent man; unit of dose of ionizing radiation that produces in man the same biologic effect as 1 roentgen of x rays or gamma rays.

Respirable suspended particles Airborne material—e.g., dusts, mists, smoke, and fumes—that is small enough (approximately 2.5 μm or less) to penetrate the lungs on inhalation. (Abbr., RSP.)

Revenue passengers Passengers who purchase tickets.

Revenue passenger mile One revenue passenger transported 1 mile.

RH See Relative humidity.

RSP See Respirable suspended particles.

Sarcoidosis A chronic disease of unknown cause characterized by widespread lesions, usually in the lungs and also in the lymph nodes, skin, liver, spleen, eyes, fingers, and parotid salivary glands.

Scheduled airline An airline that operates according to a published flight schedule specifying times, days of the week, and points between which flights are performed.

Seat hours The number of seats installed multiplied by the flight duration in hours.

Sidestream smoke Aerosol emitted into the air from a smoldering cigarette.

Smoke hood A type of protective breathing device that covers the head and face, to protect the wearer from breathing smoke and toxic fumes.

Smoke mask A type of protective breathing device that covers the mouth, to protect the wearer from breathing smoke and toxic fumes.

Strato-sphere The atmospheric region above the tropopause, having an upper limit of approximately 260,000 ft (80 km); it has very little moisture; its temperature increases with altitude.

"Stretched" aircraft An aircraft in which seating capacity has been increased beyond the designed capacity.

Total suspended particles Total mass of particles suspended in air; includes particles smaller than or equal to 10 μm. (Abbr., TSP.)

Transport category aircraft Aircraft intended for use in transportation of passengers; these aircraft must meet design, structural, and performance requirements of 14 CFR 25.

Tropopause The boundary between the troposphere and the stratosphere.

Tropo-sphere The atmospheric region in which all weather phenomena occur, from the surface of the earth up to an altitude of approximately 26,200 ft (8 km) above the poles of the earth— at midlatitudes approximately 36,000 ft (11 km) and over the equator approximately 52,500 ft (16 km); temperature steadily decreases as altitude increases.

TSP See Total suspended particles.

Type certi-fication Approval by FAA of a new aircraft design, or significant modification of an existing design, to ensure compliance with all applicable statutes and regulations.

Vapor pressure The pressure of a vapor in equilibrium with its liquid or solid form.

Ventilation The process of supplying and removing air mechanically to and from occupied spaces of an aircraft; air might or might not be conditioned.

Ventilation rate Amount of fresh air (outside air) supplied to occupants; measured in cubic feet per minute per occupant.

Wide-body aircraft An aircraft with two passenger aisles, seats for 7–11 passengers in each row (in coach), and usually a total of 200 or more seats, e.g., B-747, B-767, DC-10, L-1011, and A-300.